T0210684

Runtime Verification

Christian Colombo • Gordon J. Pace

Runtime Verification

A Hands-On Approach in Java

Christian Colombo
Department of Computer Science
Faculty of Information
and Communications Technology
University of Malta
Msida, Malta

Gordon J. Pace
Department of Computer Science
Faculty of Information
and Communications Technology
University of Malta
Msida, Malta

ISBN 978-3-031-09266-4 ISBN 978-3-031-09268-8 (eBook)
https://doi.org/10.1007/978-3-031-09268-8

This Springer imprint is published by the registered company Springer Nature Switzerland AG
The registered company address is: Gewerbestrasse 11, 6330 Cham, Switzerland

Contents

Foreword

> *In all my many years of software development, I have never seen runtime verification done as part of the testing process and now I cannot see how you could test a software product without it.*
>
> R. T. Brownrigg

The above quote is an unedited snippet from an e-mail by Richard Thomas Brownrigg. Richard is a distinguished software developer and project leader with numerous decades of experience. He wrote the e-mail to his wife, directly after participating in a one-day tutorial about Runtime Verification given by one of the authors of this book. I quote the message here, with Richard's permission, because it represents very well two truths about the topic at hand. First, Runtime Verification is not yet a mainstream technique in software development practice (at the time of writing this foreword). Second, Runtime Verification, in particular in the incarnation that Colombo and Pace put forward here, is very accessible and convincing to software developers once they are exposed to it, offering huge benefits on relatively little costs.

I am convinced that this book will have a strong impact on spreading the usage of Runtime Verification in software practice, thereby contributing greatly to the quality of the development of software systems, and indeed to the safe operation after deployment. The reason for my conviction is not simply the potential of Runtime Verification as a technology. That also. But it is very much the particular approach and style Colombo and Pace take in this book. It is not meant to barely inform the reader about the topic, but to put the reader in the position to actively use Runtime Verification in development and deployment of real software. The reader is part of a very practical journey, through exercise-centric chapters, which provide step-by-step insights into the usage of the discussed concepts and techniques.

When reading this book, what amazed me most, however, is the approach Colombo and Pace take to learning. The best way to truly understand any concept is to first invest into solving problems *without* it. Only then, after having experienced the targeted challenges, the new concept is introduced, whereafter we can *experience the empowerment* that the introduced concept provides us with. The authors follow this philosophy quite strictly and carefully. Manual monitoring is exercised before aspect-oriented programming is introduced, allowing for a better separation of concerns, and higher automation when realising monitored systems. This is, again, exercised, before guarded command specifications are introduced, allowing to specify properties that are easier to write, read, and maintain than aspects. Once more, this is put into practice before the introduction of automata allows us to, more concisely, match the (graphical) intuition of a system manoeuvring through different modes of operation. The book moves on in exactly this way to further topics: regular expressions, temporal logic, real-time monitoring, reactive monitoring, and offline verification. Time and again, the course of the text and the accompanying exercises assure that every new plateau of the toolbox is explored practically, before the next plateau offers new solutions to the problems we have already experienced.

With that, I wish the reader a great journey, empowering you with a new way of developing and deploying robust software solutions.

Gothenburg
May 2022

Wolfgang Ahrendt

Preface

Runtime verification has been around for over 20 years, with a substantial body of literature and numerous tools available. In the couple of decades we have been working within the area of runtime verification, we have witnessed and contributed to the growth of the field: the theoretical underpinnings, new verification techniques, new instrumentation strategies, development of tools, and their application in industrial case studies.

The techniques refined over these past years feel ripe for adoption in industry and yet, while one can find various theoretical overviews and academic articles, a hands-on manual which guides the reader from zero knowledge to sufficient practical knowledge required to consider its use in industry is still sorely missing. Given how industry-friendly runtime verification is, lack of such a text is surprising and regretful.

This book aims to fill this void, starting with no assumptions on the knowledge of the reader and providing exercises throughout the book through which the reader builds their own runtime verification tool. All that is required are basic programming skills and a good working knowledge of the object-oriented paradigm, ideally Java.

The book does not aim to be an exhaustive overview of the area of runtime verification. Many important research questions are still the topic of active research and discoveries, but in this book we stick to the practical underlying principles and techniques. At the end of the book we touch upon a number of more specialised areas, such as the monitoring of distributed systems and real-time systems, but such advanced topics are beyond the scope of the book. Furthermore, although we cover the pragmatic side of runtime verification, we do not delve into the theoretical foundations. However, for each chapter we do provide a reading list in the appendix for the interested reader who would like to deepen their knowledge in a particular area.

Reading guide: The first six chapters of the book should be read in order and are prerequisites for all the others that follow. Chapters 7–10 can be read independently, since each look into the use of different specification language requirements: whether automata, regular expressions, linear time temporal logic, or real-time properties. Similarly, Chapters 11–13 are largely independent, focusing on the practical implications of runtime verification: Chapters 11 and 12 are duals of each other on the spectrum of monitor intrusiveness, while Chapter 13 gives a brief overview of a number of more advanced topics, ranging from concerns of efficiency and persistence, to integration with testing and architectural considerations.

Christian Colombo
Gordon J. Pace
April 2022

Acknowledgements

The authors are indebted to the Runtime Verification community for many informal discussions about the contents of the book. We are also grateful for the opportunity to present early (and redacted) versions of the material herein as part of Escuela de Ciencias Informáticas (ECI) intensive courses organised by the University of Buenos Aires, and at the two organised schools as part of the ARVI COST Action (IC1402). We would also like to thank our runtime verification students over the years, too many to name, but whose input was crucial to improve the quality of the material.

Chapter 1
The Need for Verification

Despite the complexity of designing and implementing algorithms, the need to ensure that the systems we build work as originally intended is frequently overlooked or discounted. However, the verification of algorithms and software systems, particularly in critical domains, is crucial, and their failure can have disastrous consequences.

1.1 The Rise of Algorithms and the Need for their Correctness

Mass production is frequently hailed as a huge stride in the manufacturing industry. Instead of each individual having to learn a myriad of skills to be able to single-handedly carry out the whole production process, mass production promotes the idea of an organisational structure such that each individual produces a limited number of standard components which are then combined or assembled to create the final product. In this manner, each individual need only master the limited skillset required for the subtask he or she is responsible for, thus avoiding the need for wide-ranging skills. In the same way that the strength of a chain is dependent on its weakest link, decomposition of processes avoids that the worst of an individual's skills directly impacts the final product.

In the domain of data processing, one finds a similar revolution. Since the earliest records found, calculation was closely associated with the notion of algorithms: sequences of deterministic steps which, if followed faithfully, guarantee that a correct answer will be produced. Millennia before the birth of the computer, many such algorithms were discovered and proposed: to compute the area of a shape, to keep accounts, to calculate arithmetic over fractional numbers, to find the solution of a quadratic formula, to mention but a few. It was evident right from the beginning of this endeavour that the correctness of these algorithms is crucial. If we have an algorithmic way

C. Colombo, G. J. Pace, *Runtime Verification*,
https://doi.org/10.1007/978-3-031-09268-8_1

to deal with fractions, and we compute the taxes due by a landowner in a manner that is simple enough to be followed by a scribe trained only in the skills of basic arithmetic, then we can compute taxes efficiently. Errors by the scribes can be easily addressed through redundancy — having multiple scribes perform the same calculation and ensuring that their results match. But if the algorithm itself is incorrect, the answers will match, albeit with an incorrect value.

With certain activities, the need for precomputing the results for such algorithms for a large number of inputs was essential. Consider maritime navigation, which required complex calculations to determine the longitude of a ship from time measurements. It was not feasible to have experts aboard every ship to perform such computations every day. Instead, books of tables were produced so that a sea captain need only look up the measured inputs and have the result immediately at hand. This required performing these computations on a huge amount of data, requiring an ever increasing army of persons to perform them. The required skills, albeit limited, of the individual persons computing the results was still a bottleneck, and one way this was addressed was to decompose the algorithms into simpler and more basic units of computation, thus ensuring that less-skilled individuals would have the necessary proficiency to perform their part of the computation. For the sake of example, whereas one might have assumed the ability to multiply large numbers before, this can be decomposed in terms of repeated addition — a much simpler activity. This led to the rise of automated mechanical machines to compute results, and to Charles Babbage's dream of the *analytical engine* — a machine which could compute any algorithm expressed in terms of a number of basic activities which it 'understood'.

With increased efficiency of computation, more complex algorithms could be feasibly carried out, resulting in the development of more computationally intensive algorithms. And with this increase in the complexity and size of the algorithms, came a higher risk of errors in their design and implementation. But as long as algorithms were sparse and data was plentiful, the required proofs of correctness of the algorithms in use were few, and it was only the computation that was long and tedious. Until then, mathematicians and engineers took on the role of coming up with such algorithms and showing their correctness, but with the rise of computers in the 20th century all this changed.

User-programmable computers brought about a number of changes. The individual tedious computations, previously done by humans, could now be done much faster and much more reliably by a machine. Even more algorithms which would have been considered unfeasible before now became possible to execute. Furthermore, algorithms could now be written and executed with greater ease. Algorithms thus became more plentiful and more complex, resulting in algorithmic errors becoming ever more commonplace. To exacerbate the situation, the design and writing of algorithms no longer remained an activity exclusive to skilled mathematicians and engineers, and people with no formal training started writing code, leading to even more faulty algorithms.

Eventually, the integration of computer systems with everyday devices meant that their impact on our daily lives became even more critical. Doubtlessly, using the results of an algorithm to guide a ship across an ocean was a critical activity in itself, but when algorithms suddenly started controlling pacemakers, medical devices, military devices and weapons, vehicles, financial transactions, etc. the common person depended on their correctness on an everyday basis, and correctness became more critical than ever before.

1.2 What is at Stake?

As we have just argued, the outcome of incorrect algorithms goes beyond incorrect results. Software is increasingly being used in critical systems, and the cost of bugs (financial or otherwise) in such system can be prohibitive, making the need for correctness ever more important. The mathematical algorithms developed and used as far back as the 15th and 16th century to compute navigational tables to enable the calculation of a ship's location are an early example of such a system of a critical nature. Errors in the algorithms would have led to incorrect tables which, in turn, would have led to ships being lost at sea.

Today, with software automating so many everyday processes, business or otherwise, the potential losses are huge, with many documented instances of such cases. The negative impact of bugs can take different forms.

Financial loss. One finds various instances of bugs which led to large financial losses. Consider Knight Capital Group, a financial services company, enduring a loss of 460 million US dollars in less than an hour due to a bug in a trading algorithm. A payroll system developed by IBM Australia for Queensland Health led to tens of thousands of payroll mistakes, resulting in losses of 1.25 billion Australian dollars. Parity Technologies had a bug in a smart contract (a form of software) deployed on the Ethereum blockchain, which led to a loss of the equivalent of over one million ether cryptocurrency (equivalent to 300 million US dollars at the time of the loss) endured by its users. The bug set the smart contract in a state in which the cryptocurrency stored within it could not be withdrawn under any circumstances[1].

Credibility loss. Together with financial loss, software failures also result in the companies involved facing loss of credibility and trust from clients.

[1] In a smart contract running on a blockchain one cannot simply manipulate the code, the state of the software or the values of the balances to restore it to a valid state. This is unlike more traditional setups such as a database on a server where such rectification is possible.

Consider an IT failure of a British Airways system in 2017 which resulted in the cancellation of over 700 flights and 75,000 stranded passengers. Another well-publicised case was a bug in the Google+ social network which led to the private data of half a million accounts being exposed. Google announced the closing down of the social network shortly after the bug was exposed in 2018. The impact of such a loss of client trust can have a longer term impact on the company providing the software well beyond the immediate loss of a financial nature.

Operational failure. With certain systems, software failure results in the operations on which the software depends failing altogether. Consider the 1995 loss of the European Space Agency's Ariane 5 rocket which was lost less than a minute after takeoff due to failure of the on-board guidance software. Another case was an hours-long blackout across a number of US and Canadian states in 2003 which was the result of a software bug in a monitoring system which stopped the electricity load from being redistributed from overloaded transmission lines.

Loss of life. The most serious of software bugs may result in the loss of human life. In 2018, Fiat Chrysler had to recall almost 5 million cars due to a bug in their cruise-control software which sometimes stopped the driver from being able to deactivate cruise-control. Although not recorded to have led to loss of life, such a bug could have easily resulted in a car crash. Health is another domain in which faulty systems can result in deaths. In 2014, Carefusion had to recall one of their models of infusion pumps (which deliver drugs in an automated way to patients) due to a bug which could lead to drugs being infused earlier or later than intended, possibly with lethal results.

As these cases show, for many software systems, failure is not a viable option. Losses of varying severity can result, showing that the responsibility to deliver software that works correctly goes beyond being merely desirable (a better user experience) to being a necessity (from a business or human-life perspective).

1.3 Why is Software Failure so Common?

The number of problems we experience when using software and the number of media reports on high-profile IT failures far outweigh our experience with more traditional engineering artifacts. Why do we comparatively rarely hear of mechanical failure of machines due to bad engineering, collapsing bridges or falling lifts? This imbalance can be partly justified by referring to frequency of interaction — nowadays, we interact with software in a practically continuous manner and we are thus more likely to experience failure in such systems even if the percentage of failing software systems may not be that much higher than for other engineered systems. But this argument still fails to justify the huge disparity.

Various reasons have been proposed explaining why software systems fail so frequently. One is that of the complexity inherent in the design, that the state space of a digital system is so large that it is difficult for a human team to account for all eventualities. Another related argument is that digital systems demonstrate a degree of instability not frequently found in analogue systems. In most engineered analogue systems, a small change in the input (e.g. providing just a little extra force) will result in a small change in the output (e.g. a small increase in final velocity), and traditional engineers are trained to use rules-of-thumb to keep away from discontinuities (e.g. a force which will break the object rather than accelerate it faster). The results achieved when testing such a device on a particular input gives valuable information on how it will behave not only with that particular input but also with similar ones, thus allowing for more effective evaluation. In contrast, digital systems do not exhibit this property. Toggling one bit of a binary number (the smallest change one could make) may yield a completely different value, particularly if you switch the value of the most significant bit. Analysing such systems for correctness is thus much more challenging.

No matter the degree of difficulty inherent in designing and building digital systems, errors can certainly be reduced by adopting good engineering practices. Bugs and failures can be due to a combination of bad practice and bad tools: (i) bad processes: including disorganised engineering processes, insufficient quality assurance safeguards, lack of team organisation or cohesion, lack of responsibility management, etc.; (ii) bad tools: including bad choice of programming language or framework for the task at hand, problems with underlying platforms, etc.; and (iii) bad code: resulting from bad design, bad structure, bad data organisation, lack of compositionality, etc.

However, whatever the source of bugs and erroneous code, there is the need for software analysis techniques to be able to identify, remove or at least limit the impact of such bugs. One approach is software verification, where the desirable and/or undesirable software behaviour is specified in terms of a specification language, and a corresponding verification process takes the specification as input, and performs analysis on the system description and/or actual behaviour (see Figure 1.1).

This book is about one form of software verification called *runtime verification*. This verification technique advocates the building of monitoring software whose role is to observe the main system and ensure it works as expected, intervening appropriately when errors are observed, with the assumption that such monitors are substantially simpler to build correctly than the system itself (an assumption which we know to be true in practice). However, we will be delving deeper into what runtime verification is, and how it can be used in practice in the coming chapters. But before we do so, we will identify a number of prerequisite questions for verification.

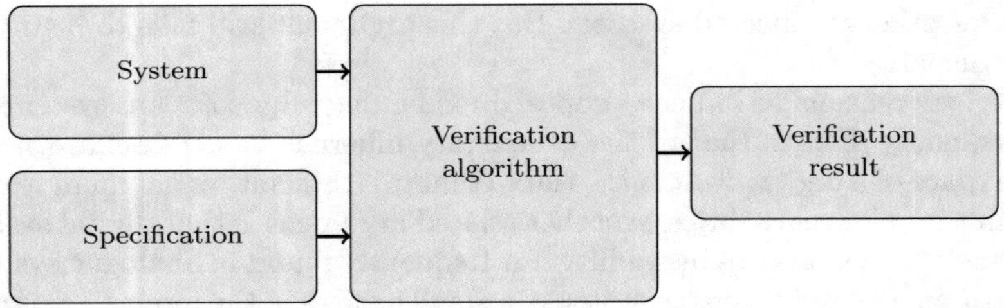

Fig. 1.1 Software verification

1.4 Some Pertinent Questions

Before delving deeper into runtime verification, it is worth asking a number of questions in order to clarify what and how we would like to verify our systems. The rest of the book will then be spent answering these questions.

What are the sort of questions we want answered about systems? One can identify a number of sanity checks to capture problems common to all systems, such as memory leaks and incorrectly used language structures or bad use of standard library functionality. At the simplest level, one finds linters, intended to highlight basic bugs and stylistic errors such as unused variables or uninitialised objects. Whilst linters typically stop at a syntactic analysis of the code, more complex analysis tools exist, looking at the flow of the logic to try to refine the capturing of such problems. Whilst such analysis is important for the smooth running of the software, we also need to ensure that it is fit for purpose — that the implementation of the business logic (the real-life processes it is automating) is correct. In this case, we need to be able to specify what the code is meant to do, which is specific to each and every system: a payroll system must ensure that salaries are computed correctly and issued on time, whilst a financial transaction system must ensure that only the sender's and recipient's account balances are changed when a transfer of funds occurs. Such properties would have to be written specifically for the system in question. Such properties may talk about events at the real-life level e.g. talking about a financial transfer between a sender and a receiver, though in practice, to perform verification, they would have to be related to the actual code e.g. an outgoing financial transfer may correspond to a call to either the function `makeTransfer()` or `performStandingOrder()`. This leads us on to the next question.

How do we specify the properties we want to verify? Properties we may want to verify may come in various forms, from (i) very code-centric ones e.g. an invariant of the form *"The sum of the balance of an account and the account's overdraft allowance may never be negative"*, or a pre-/postcondition of the form *"Whenever function* `withdraw()` *is called on an account with a parameter no larger than that account's balance, if the function successfully* *terminates, then the balance of the account must be reduced by the amount specified in the parameter"*, to (ii) properties about the system's behaviour possibly spanning across its lifetime e.g. *"Once an account is greylisted, it loses its gold status and may not regain it until it is whitelisted again for an uninterrupted period of one month"*. Clearly, the choice of the form of specification and the complexity of the statements used to write specifications about the system's behaviour will have an impact on how difficult or expensive verification will be to perform.

How do we verify these systems? Finally, once we have specifications, we have to see what techniques can be used to ensure that the system-under-scrutiny works as expected. There are various such approaches, including testing with which most developers are familiar, to runtime verification, which this book is about, and static verification. We will be briefly explaining these approaches in the next chapter in order to give a better context to runtime verification. It is worth keeping in mind that different techniques may provide different guarantees or address different specification languages. We will go into this in more detail in the next chapter.

Now that we have established the importance of verification, and identified the main issues which will arise when deciding how to verify our systems, we will move on to explaining what runtime verification is, and how it relates to other verification techniques. In the rest of the book we will also be presenting different approaches to runtime verification, which further answers these questions with a specific focus on runtime verification. Furthermore, we will be implementing these techniques into our tool, and by the end of the book we will have built a complete runtime verification tool for Java systems.

Chapter 2
What is Runtime Verification

Once the risk of incorrect code is deemed sufficiently high for a particular software system, one has to decide on how best to verify it in order to mitigate such a risk. Testing has become a given in software engineering, but its limitations are well-known and acknowledged. The question is how one can achieve more certainty in a practical manner. In this chapter we introduce the notion of runtime verification as one such approach.

2.1 Testing and Exhaustive Analysis

Testing is by far the most commonly used technique to identify errors in code. By defining test cases which drive the system and conditions (known as oracles) to decide whether the system deviates from the expected behaviour, we can easily and efficiently evaluate whether the underlying system is working correctly for the identified scenarios tested. The relative ease with which tests can be written and the validation process efficiency led to testing being virtually universally adopted throughout the software industry. However, notwithstanding the quality gains derived from testing, clearly, testing may miss certain bugs. The problem is, as the computer scientist Edsger Dijkstra aptly put it, *"program testing can be used to show the presence of bugs, but never to show their absence"*.

One problem is when to stop adding new test cases. Given that testing all possible scenarios is impossible for most systems, we would like to stop testing when we believe that the written test cases cover all different possibilities, e.g. if we have reason to believe that *the system working correctly with a particular positive input implies that it will work correctly with any positive number input*, then it would suffice to test with input 17 (or any other positive value). The problem is that the notion of coverage is a difficult one to define and measure in an effective manner. One can talk about different types of coverage: data coverage (what percentage of possible values

C. Colombo, G. J. Pace, *Runtime Verification*,
https://doi.org/10.1007/978-3-031-09268-8_2

have been covered by the tests), control-flow coverage (whether tests went through all instructions in the program) or property coverage (whether all the logic of the properties has been covered by the tests), making the notion of sufficient coverage far from straightforward.

If the system is simple enough to be able to test all possible paths, we would be able to decisively conclude whether or not the system is correct with respect to the properties we are testing for. Alas, with most systems, the number of possible test cases is infinite or so large that such a complete approach would not work. However, techniques have been developed for exhaustive analysis using intelligent approaches in order to make the analysis feasible without having to test each path individually[1]. Such techniques — static analysis, model checking, symbolic execution, etc. — have been shown to be useful for the verification of certain real world systems. However, in general, the approaches still require expert interaction and typically do not scale up to provide a push button solution to the verification of software of the scale of real-world systems.

In an ideal world, we would like to ensure that our systems never fail, but testing leaves out paths which may occur in practice and exhaustive analysis does not scale up. This is where runtime verification comes in.

2.2 What is Runtime Verification?

Based on the discussion in the previous section, we can make a number of observations. The first is that checking a whole system for all possible inputs and execution paths is computationally expensive, but checking a single path is typically not. This is why checking a single test case is computationally tractable, whereas exhaustive verification is not. The second observation is that obtaining a sufficiently complete set of paths for testing is tough because no matter how many paths we check, there will still be others which have not been tested. If such untested paths occur at runtime, we have no guarantee that they will work correctly. Based on these two observations, we can identify a way forward: Once the system is deployed, we continue checking its behaviour to ensure that we are aware whenever an execution path followed by the live system violates a property; if a violation is detected, we can choose to act upon it, possibly fixing the system or even stopping it, depending on the severity of the violation. Since we are checking a *single path* at runtime, we know that the verification process is typically not computationally expensive, and furthermore, since we check the path observed *at runtime*, we

[1] You can think of these algorithms as attempting to generate a mathematical proof that the program satisfies the property. Their approach is similar to the way that in mathematics one can prove a statement about numbers (think of Pythagoras' theorem) without having to try out each value individually.

can be sure to be able to intervene as soon as, if not just before, the system reaches a point of failure.

What we have just described is runtime verification. To be more precise, there are three concepts described in the previous paragraph:

- **Runtime monitoring.** "Once the system is deployed, we continue checking the system behaviour".
- **Runtime verification.** "Ensure that we are aware whenever an execution path followed by the live system violates a property".
- **Runtime enforcement, recovery or reparation[2].** If the system behaviour violates a property or is about to do so, we can react accordingly, "possibly fixing the system or even stopping it".

Throughout the rest of the book we will be using these terms as described above[3].

Consider, for example, the specification which states that *"A user may only download a file from the moment they log in until they log out"*. Monitoring code would be added to the system to identify when a user is logging in, logging out, and downloading a file. The verification code would then have do the following: (i) for each user, add a variable set to *true* upon observing a login event, and *false* upon observing a logout event, thus remembering whether or not a user is logged in; and (ii) whenever a user is about to download a file, check whether that user is logged by using this information. If the variable is set to *false* (i.e. the user is not logged in), the runtime verifier reports a violation. Finally, code can be added to intervene whenever such a violation is reported. This would depend on how the system designers would want to react in such an eventuality: stopping the download from happening, blocking the user (if we suspect that the user is trying to bypass system security), or even simply notifying the developers of the unexpected behaviour and allowing the download to happen (if such a violation is not considered serious enough to disrupt the user experience).

As a more complex example, consider the property: *"Once an account is greylisted it loses its gold status, and may not regain it until it is whitelisted again for a continuous period of at least one month"*. Additional monitoring code is injected into the system so that we can capture calls to functions which greylist, whitelist, and give or remove gold status to an account. Then, verification code is added to ensure that (i) after an account is greylisted, it does not have gold status; (ii) remember the most recent timestamp when an account became whitelisted; (iii) whenever a previously greylisted account is given gold status, the account must be whitelisted, and must have been whitelisted for at least one month, which can be checked from the timestamp

[2] We use all three terms since sometimes we can stop bad behaviour from happening, whereas in others we can only reset the system back to a valid state or perform reparatory action to make up for the fact that a violation occurred.

[3] In the literature, one frequently finds the term *runtime verification* used to refer to all these approaches, hence the name of this book.

of the latest whitelisting event. Finally, recovery code can be added in case this property is violated.

The good thing with this runtime approach is that we can be sure that the system is always aware of a violation. Furthermore, since checking a single trace is typically not so computationally expensive, the approach scales up to real systems and as we shall see in the rest of the book, it is easy to adopt this approach to be used in real-world software projects.

The downside is that we can still never say that the system is correct, and we do not remove bugs, but rather react to them, possibly stopping their consequences. Furthermore, since the analysis is meant to take place at runtime, it induces additional overheads in terms of execution time and memory on the system in order to run the monitoring and verification code. In practice, these overheads are typically low, but they may still interfere with the underlying system, and in particular cases, they can be prohibitive.

In the rest of this chapter, we will be looking at a number of important issues arising if we want to use runtime monitoring, verification, and recovery in our systems. We will then look at how a runtime verification and recovery tool can be built in the following chapters.

2.3 Programming Runtime Monitors and Verifiers

Runtime verification is thus the process of adding monitoring, verification and possibly recovery code to the main system logic. The most straightforward way in which this can be achieved is by adding such code to the main system as it is developed. However, such an approach comes with a number of shortcomings:

- Making no distinction between system logic and monitoring, verification and recovery logic results in an increase in complexity of the main system code. Consider if we want to detect whenever the function `download()` is called before function `login()` (e.g. we may believe that the user interface should never allow this, and therefore, if it occurs it would indicate that the system must be malfunctioning). In order to perform this check, we would have to add a Boolean flag to keep track whether `login()` has been called. The flag would be initialised to *false*, updated to *true* when `login()` is called, and checked to be *true* whenever `download()` is called. This extra data and logic becomes intermingled with the main system code. You may argue that the code is so simple that it is highly unlikely to have a bug, or to interfere with the system code, causing new problems. However, consider if we had to check for a more complex condition such as when `login()` is called when `badPassword()` was invoked at least three times in the past half hour. The monitoring and verification code is substantially more complex now, and this is clearly undesirable.

- If the code for monitoring and verification is inlined with the main system logic, most probably the development team will also be responsible for handling the monitoring and verification concerns. The developers will likely perceive parts of the additional code required for verification as a replica of internal sanity checks that are already being built into the system. Referring back to the previous examples, if the user interface already has a flag to remember whether a user is logged in, why not just reuse that flag, or if the system already has a way of counting how many bad logins were received in the past half hour, why not use that information? The problem is that the aim of the new logic is to check that the system is working correctly. If there is a problem with these bookkeeping checks kept by the system, the verification code will not be able to notice anything untoward.
- Maintenance of inlined monitors and verification code across different versions of the system can be challenging, since the code of a single property can be spread across different parts of the system. Modifying any of these parts or adding new parts which may affect the property would require changes and additions to the monitoring and verification code, and may easily end up out of sync. By separating the specification from the system, it can typically be reused across system versions in a more straightforward manner.

Separating development from quality assurance is considered good practice. Apart from avoiding the problem of code reused for monitoring and verification, having a separate team interpret the requirements and develop monitors and verification code is beneficial. In order to ensure such decoupling of functionality and verification code, runtime verification advocates for a separation of concerns between the system-under-scrutiny, and the monitoring and verification code, writing them separately, and using a tool to weave the two together. Having said that, in this book we will start by showing how such monitoring and verification code can be inlined, after which we will develop techniques to decouple it from the system.

The way a runtime verification tool is used is shown in Figure 2.1. The runtime verification tool processes the system code and the specification (i.e. the monitoring and verification instructions) and generates an updated version of the system with additional code to monitor, verify, and handle recovery with respect to the properties written in the specification.

2.4 Choices in Runtime Verification

The process of runtime verification as discussed in the previous section still leaves a number of choices open, which we will discuss.

How do we express the properties? In the previous chapter we have discussed that there are different types of properties we may be interested

Fig. 2.1 Runtime verification tool-flow

in specifying. One distinction can be made between code-centric properties referring to particular points in the code and temporal ones spanning system behaviour across the code. The former correspond to programming assertions, and speak about properties that should hold at particular points in the execution of a program e.g. the moment a transfer is about to be made, the sender must have enough balance and may not be blacklisted. The latter type of properties speak about how the system evolves and the order in which things should happen e.g. downloads may only happen between a login and a logout. Point properties are frequently used by developers and are tied tightly to the code itself. Temporal properties, on the other hand, do not always belong to a particular piece of code and relate to the system's behaviour as a whole. Most work in runtime verification focusses on the latter, and one of the important choices we have to start by making is how to specify such properties. In the course of this book we will be looking at different formalisms which can be used, ranging from graphical formalisms to temporal logics which can provide a more succinct way of expressing certain properties at the cost of quality assurance engineers having to learn a new formalism.

How do we verify properties? Earlier in this chapter we outlined how one could verify the temporal property: *"Once an account is greylisted it loses its gold status, and may not regain until it until it is whitelisted again for a continuous period of at least one month"*. The proposed algorithm to capture this property is far from a straightforward implementation of the property, and one could ask whether it correctly implements the property. As we adopt richer means of expressing temporal properties, the translation to verification code is not always obvious. In later parts of the book we will explore techniques which can be used to translate properties into verification code automatically.

How do we instrument monitoring and verification code? The process of adding code to the original system is called *instrumentation*. Typically, the monitoring process is instrumented such that the only code added is the generation of events which are consumed by the rest of the monitoring and verification which is kept separate. As we shall see in the coming chapters,

one can either require the developers to manually instrument the code for the event generation, or use tools to instrument such code automatically.
How does the verification code communicate with the main system? Until now, we were tacitly assuming that the monitoring and verification code was running in lockstep with the system. That is, whenever the system reaches a point of interest e.g. a call to a function referred to in the specification, it does not proceed before the verifier gives it its go-ahead. In this manner, violations can be captured immediately and acted upon before they cause further problems. However, this process slows down the system, particularly if complex verification tasks are taking place. One other option is to perform verification asynchronously with the main system, which will simply log an event and continues unhindered. Verification may then take place independently e.g. on a separate machine thus not slowing the system down (beyond the event generation). In this case, however, violations may be discovered late and may have already wreaked havoc in the system. In the end, it is a question of weighing the gains in performance against the consequences of discovering a violation late.
How do we deal with recovery when a problem is found? Handling system recovery is one of the hardest aspects of runtime verification. Apart from being very system specific — requiring intimate knowledge of how the system works — recovery would typically require changing the system's state e.g. updating the status of a user. However, such interference can easily lead to breaking system-wide invariants. Consider the property we saw earlier which required (amongst other things) that greylisting a user should strip them of their gold status. If greylisting is only performed in one or two places in the code, the developers may have chosen to remove the gold status just after the greylisting procedure is called, rather than include it as part of the greylisting procedure itself. Now consider the case where another property is also used and whose violation would indicate that the user may be trying to bypass security measures. As a reparation for this second property, the quality assurance engineers may decide to greylist the user, thereby stopping them from performing transfers. But in doing so, they may not realise that they also have to, separately, take away the user's gold status. This reparation would thus inadvertently introduce a violation of the first property.

2.5 Conclusions

In this chapter we have introduced the notion of runtime verification, explaining how it complements testing and scales up better than exhaustive verification. There are various issues to consider when implementing a runtime verification tool, issues which will be examined in more detail in the coming chapters as we progressively implement our own runtime verification tool.

Chapter 3
FiTS: A Financial Transaction System

Until this point in this book, we have used various examples to illustrate concepts. It helps, however, to have a recurring non-trivial use case with a number of properties to use throughout the book. This helps us avoid having to explain a different use case and the reader having to dedicate time afresh to understand it each and every time. In this chapter we will present such a use case which will be used in the rest of the book. Financial Transaction System (FiTS) is a cut-down version of a financial transaction system to enable experimenting with runtime verification techniques on its behaviour. The system has been kept reasonably simple to ensure that the focus is on the verification techniques and not on understanding the underlying system. In this chapter, we present basic information about FiTS to allow us to use it for our examples and exercises throughout the book.

3.1 Understanding the Structure of FiTS

Financial Transaction System (FiTS) is a bare-bones system mocking the behaviour of a financial transaction system — a virtual banking system in which users may open accounts across which they may perform money transfers. FiTS is built around a number of concepts such as users, accounts and sessions. These terms are defined below:

Administrator: The administrator of FiTS has more rights than normal clients to be able to perform certain actions such as approving the opening of an account, enabling a new user and registering new users.

User: In FiTS, clients or users have access to a number of actions to access their (money) accounts which they have registered for on FiTS— actions which they may invoke through an online interface. Information about

C. Colombo, G. J. Pace, *Runtime Verification*,
https://doi.org/10.1007/978-3-031-09268-8_3

each registered user is stored in a database,[1] including information such as the client's name, country of origin and user classification information:

User type: Clients fall under three types or categories: Gold, silver or normal. The user type determines charges which he or she will incur when transferring funds using FiTS.

User status: Each user is either white-, grey- or blacklisted. Greylisted and blacklisted users may be stopped from performing certain actions.

User mode: A user may be in one of a number of modes: (i) *enabled* users are allowed to log in and make transfers; (ii) users start off as *disabled* (before being activated by the administrator) and may also end up in this mode if they are discovered to have engaged in fraudulent behaviour; (iii) users may temporarily freeze their user account — *frozen* users may not perform money transfers, thus ensuring better security for the user if they do not plan to use FiTS for a while. At any point, a frozen user may reset their mode back to active.

Account: Each user account may be associated with a number of money accounts belonging to him or her. Users may request the creation of new accounts or close down existing ones they own. An account may be in one of four states: *requested*, *open*, *closed* or *rejected*.

Session: An existing user may open a login session on FiTS to make transfers or manage their accounts. A user may have multiple sessions open at the same time.

The FiTS code primarily consists of the backend code which handles the underlying database and content management in the system. Only certain backend methods can be accessed by the front end of the system which the administrator and users have access to. Although the front end typically ensures that the behaviour of the users is limited e.g. not allowing a blacklisted user to make a transfer, the backend has been built with the intention of double-checking that such violations do not occur.

3.2 FiTS Repository

FiTS can be downloaded from `https://github.com/ccol002/rv-book`. The code repository is organised in folders, the names of which are identified in the book exercises. The code is typically organised per chapter with code relevant to that chapter being found in a folder called `chapternn`, where `nn` is the chapter number. For instance, you will find the code for the clean system as discussed in this chapter in the folder `chapter03`. Some chapters are organised further into parts, with the folder names mentioned in the text. The code for

[1] For simplicity, FiTS only stores the data in memory, and does not keep a database in secondary storage.

some of the chapters is split into parts — for instance Chapter 5 has two parts which will be in separate folders `chapter05` and `chapter05-inlined`. The references to these folder names appear in the exercises and examples, and shown as chapter05-inlined.

The code augmented with solutions to the exercises can be found in a folder with the same name as the exercises, but appended with `-solutions`. For instance, the solutions to this chapter can be found in the folder named `chapter03-solutions` while the solutions of `chapter05-inlined` can be found in `chapter05-inlined-solutions`.

Since much of the content of this book is incremental, in that many of the chapters assume that you have already solved the exercises from previous chapters, the folder for a particular chapter will include the solution code for any such dependencies. If you prefer to use the code you developed, you will have to copy the relevant files to the new chapter. For example, certain exercises in Chapter 7 require you to have solved the exercises from Chapter 6, so the code in those Chapter 7 folders will include solutions to the relevant Chapter 6 exercises.

Exercise 3.1

Download the FiTS repository, and open the version of FiTS related to this chapter in your preferred IDE.

Have a look at the code, where you will find a readme file referring you to this chapter's solutions folder.

chapter03

3.3 The Modules

The FiTS code provided primarily consists of the backend code which handles the user and accounts information database behind the system. From this backend, only certain methods can be accessed by the frontend of the system which the administrator and the user have access to.[2] Although the frontend typically ensures that the behaviour of the users is limited e.g. by not giving a blacklisted user the option to make a transfer, the backend was built with the intention of double-checking that such violations do not occur.

Below are descriptions of the main FiTS modules and the methods of interest:

[2] In order to keep the code simple, we do not include actual code for a user interface frontend, but rather provide functions which the frontend would call in order to access the underlying functionality.

- `TransactionSystem`: The FiTS as a whole is stored as an instance of `TransactionSystem`, with the following methods:
  - `void setup()`
    Resets the transaction system to an initial state.
  - `FrontEnd getFrontEnd()`
    `BackEnd getBackEnd()`
    Getters to obtain access to the frontend and backend of FiTS respectively. The structure of `FrontEnd` and `BackEnd` is documented below.

- `FrontEnd`: The FiTS frontend
  This module contains the only methods over which the user interface for the administrator (or automated backend tools with administrative rights) and the clients may interact with the backend. All method names are prefixed with either `ADMIN` or `USER` to indicate who can access them.[3]
  The methods which the administrators may execute are:
  - `void ADMIN_initialise()`
    Initialises the transaction system.
  - `void ADMIN_reconcile()`
    Reconciles the accounts to ensure that the balances stored in the database are correct.
  - `Integer ADMIN_createUser(String name, String country)`
    Creates a new user with the given name (`name`) and country of residence (`country`), returning the new user's unique id.
  - `void ADMIN_disableUser(Integer uid)`
    `void ADMIN_activateUser(Integer uid)`
    Disables or activates a user with the given user id (`uid`).
  - `void ADMIN_blacklistUser(Integer uid)`
    `void ADMIN_greylistUser(Integer uid)`
    `void ADMIN_whitelistUser(Integer uid)`
    Black, grey or whitelists the user with the given user id (`uid`).
  - `void ADMIN_makeGoldUser(Integer uid)`
    `void ADMIN_makeSilverUser(Integer uid)`
    `void ADMIN_makeNormalUser(Integer uid)`
    Updates the type of the user with the given user id (`uid`) to be gold, silver, or normal.
  - `void ADMIN_approveOpenAccount(Integer uid, String accnum)`
    `void ADMIN_rejectOpenAccount(Integer uid, String accnum)`
    Approves or rejects the opening of an account with the given number (`accnum`) and owned by the user with the given id (`uid`). The user should have previously requested the opening of such an account.

[3] The code included in FiTS does not handle user authentication, which is assumed to be in a separate layer which would handle access control.

The methods which the users may execute are the following. All the methods include the user id of the user executing the command (`uid`).

- `Integer USER_login(Integer uid)`
  `void USER_logout(Integer uid, Integer sid)`
  Logs in the user into the system, returning the new session's id and logs out the user of the session with the given id (`sid`) respectively.
- `Boolean USER_freezeUser(Integer uid)`
  `Boolean USER_unfreezeUser(Integer uid, Integer sid)`
  Freezes and unfreezes the user with id `uid` in session with id `sid`, returning whether the action was successful.
- `String USER_requestOpenAccount(Integer uid, Integer sid)`
  Requests the opening of a new bank account owned by the user with id `uid` in session with id `sid`, returning the account number that will be assigned if approved.
- `void USER_closeAccount(Integer uid, Integer sid, String accnum)`
  Closes an existing bank account with account number `accnum`, owned by the user `uid`, and requested from session `sid`.
- `void USER_depositFromExternal(Integer uid, Integer sid, String accnumDst, Double amount)`
  Deposits money (`amount`) from an external source (e.g. from a credit card) to an account with account number `accnumDst`, owned by user with id `uid`, and from session with id `sid`.
- `Boolean USER_payToExternal(Integer uid, Integer sid, String accnumSrc, Double amount)`
  Sends money (`amount`) to an external money account (e.g. pay a bill) from the account with the account number `accnumSrc` — charges are applied and also taken from the user's account. The method returns whether the transfer was successful.
- `Boolean USER_transferToOtherAccount(Integer uidSrc, Integer sidSrc, String accnumSrc, Integer uidDst, String accnumDst, Double amount)`
  Transfers money (`amount`) from a user's account (with user id `uidSrc` and account number `accnumSrc`) to another user's account (with user id `uidDst` and account number `accnumDst`) — charges are applied and also taken from the sender's account. The method returns whether the transfer was successful.
- `Boolean USER_transferOwnAccounts(Integer uid, String accnumSrc, String accnumDst, Double amount)`
  Transfers money (`amount`) across accounts owned by the same user — from account with number `accnumSrc` to account `accnumDst`. No charges are applied in this case. The method returns whether the transfer was successful.

- `BackEnd`: The backend of the transaction system
  The underlying functionality of FiTS is given in this module. Although some of the methods are similar to the ones provided in the interface class, the ones in this class will access or change the actual state of the transaction system, and the ones in the interface class correspond to functionality being requested by a user or administrator. The main methods of interest[4] in this module are:

  - `void initialise()`
    Initialises the transaction system. For the sake of verification, the system is started in a predetermined initial state with a number of users and accounts.
  - `Integer addUser(String name, String country)`
    Creates a new user with the given name (`name`) and country of residence (`country`), returning the new user's id.
  - `UserInfo getUserInfo(Integer uid)`
    Given an existing user's id (`uid`) returns that user's information. The structure of `UserInfo` is documented below.

  Note the difference, for instance, between the method `addUser` which adds the user to the database, and the method `ADMIN_createUser` in the module `FrontEnd` which is the administrator's request to create the user and which would, in turn, call the backend function. Similar distinctions are to be seen in other methods.
- `UserInfo`: The users' information database
  The `UserInfo` object stores the state of a particular user. Important methods are the following:

  - `void makeGoldUser()`
    `void makeSilverUser()`
    `void makeNormalUser()`
    Sets the user's type to gold, silver or normal accordingly.
  - `void makeBlacklisted()`
    `void makeGreylisted()`
    `void makeWhitelisted()`
    Sets the user's status to blacklisted, greylisted or whitelisted respectively.
  - `void makeActive()`
    `void makeFrozen()`
    `void makeDisabled()`
    Sets the user's mode to active, frozen or disabled respectively.
  - `Integer openSession()`
    Opens a new session for the user, returning the session id.

[4] Other methods are defined and documented in the module, but these are the only ones relevant to the verification process.

- `void closeSession(Integer sid)`
  Closes the session with the given session id `sid`.
- `ArrayList<UserSession> getSessions()`
  Returns all active user sessions. The structure of the `UserSession` class is explained later.
- `UserSession getSession(Integer sid)`
  Returns a specific active user session identified by session id `sid`.
- `String createAccount()`
  Creates a money account owned by the user, returning the account number.
- `void deleteAccount(String accnum)`
  Deletes the money account with account number `accnum`.
- `ArrayList<BankAccount> getAccounts()`
  Returns all bank accounts owned by the user. You will find the structure of `BankAccount` is given below.
- `BankAccount getAccount(String accnum)`
  Returns the bank account with account number `accnum` owned by the user.

• `BankAccount`: Bank accounts
The `BankAccount` object stores the state of a bank (money) account. Bank accounts are designed to be owned by a single user. Verification-relevant methods are the following:

- `Double getBalance()`
  Returns the current balance of the account.
- `void withdraw(Double amount)`
  `void deposit(Double amount)`
  Withdraws (respectively deposits) the requested sum (`amount`) from (respectively to) the account.
- `Integer getOwner()`
  Returns the user id of the owner of the account.

• `UserSession`: Login sessions
The `UserSession` object stores the state of a login session and a log with the actions carried out in that session. Every session is associated with a single user who may, however, have multiple open sessions. Verification-relevant methods are the following:

- `void openSession()`
  `void closeSession()`
  Methods to open and close a session. Note that just creating a new `UserSession` object will allocate a session id to the session but will not open the session automatically.
- `void log(String msg)`
  Adds the given message to the log of the session.

– `String getLog()`
  Returns the logged messages of the session as a string.

Exercise 3.2

Take some time to familiarise yourself with the code provided in FiTS, corresponding to the descriptions of the modules given in this section.

chapter03

3.4 Is FiTS Fit for Purpose?

Given that a financial transaction system can handle transfers of large amounts of assets, it is crucial to ensure that FiTS works as envisaged. In order to do so, we would need a specification indicating how FiTS should function.

We achieve this by specifying a number of sample properties[5] which FiTS would be expected to satisfy, and which will be used in our verification efforts in the rest of the book. Needless to say, the properties provided are not comprehensive, and in a real-world scenario, one would have a more complete set of properties to be used for verification.

The properties we will be using as our specification are the following: [6]

1. Only users based in Argentina can be gold users.
2. The transaction system must be initialised before any user logs in.
3. No account may end up with a negative balance after being accessed.
4. A bank account approved by the administrator may not have the same account number as any other bank account already existing in the system.
5. Once a user is disabled, he or she may not withdraw from an account until the administrator enables them again.
6. Once greylisted, a user must perform at least three incoming transfers before being whitelisted.
7. No user may request more than 10 new accounts in a single session.

[5] For software developers, the word "property" usually refers to an object's characteristic such as an attribute. However, in the context of software verification, a formal specification is usually given in terms of a list of properties. Therefore a property is just a manageable chunk e.g. a formula or automaton, of a system's formal specification.

[6] You may assume that that FiTS is deployed in Argentina, and for compliance reasons, certain offers made to gold users cannot be offered to users residing outside the country.

8. The administrator must reconcile accounts every 1000 attempted outgoing external money transfers or an aggregate total of one million dollars in attempted outgoing external transfers (attempted transfers include transfers requested which never took place due to lack of funds).[7]
9. A user may not have more than 3 active sessions at any point in time.
10. Logging can only be made to an active session (i.e. between a login and a logout).
11. A session should not be opened in the first ten seconds immediately after system initialisation.
12. Once a blacklisted user is whitelisted, they may not perform any single external transfer worth more than $100 for 12 hours.
13. A user may not have more than three accounts created within any 24 hour period.
14. An administrator must reconcile accounts within 5 minutes of initialisation.
15. A new account must be approved or rejected by an administrator within 24 hours of its creation.
16. A session is always closed within 15 minutes of user inactivity.

It is worth noting that the English description of these properties leaves much to be desired. For instance, Property 2 says that *"The transaction system must be initialised before any user logs in"*. However, it can be argued that it is unclear whether it should be taken to mean that the `initialise()` method in the `BackEnd` should be called, or whether the requirement is for the administrator to initialise the system through `ADMIN_initialise()` in the `FrontEnd` module. As we start specifying these properties in a more precise and formal manner, we will have to take decisions on the meaning of the properties. In the real world, we would revert back to whoever carried out the requirements capturing to be more specific. When reading this book you will not have such a luxury, and in most cases, it will boil down to a question of interpretation. For instance, in the case of Property 2, one can argue that since the property makes no reference to the administrator, and since other code in the future may support other means of initialising the underlying system, it is the `initialise()` method in the `BackEnd` that will be required to have been called.

In fact, the code provided for FiTS is *not* fit for purpose, and allows for scenarios in which these properties are violated. The existence of these bugs will allow us to validate our verification tools using FiTS.

[7] Although the property states what an administrator should do, typically the system would trigger such an action on behalf of an administrator.

Exercise 3.3

Look through the code and try to find ways in which FiTS violates some of these properties. Do *not* fix these bugs, since we will be using them to check that our verification tool is working.

chapter03

3.5 The Scenarios and their Execution

The `Scenarios` module provides code to drive FiTS along particular execution paths.[8] The module comes with two such paths for each property, one violating the property and one which does not. The two main methods provided are:

```
static void runViolatingScenarioForProperty(Integer n)
static void runNonViolatingScenarioForProperty(Integer n)
```

These methods allow the running of individual tests to see the results. Information about which test is being run is displayed in the console. It is worth noting that some of the properties require a substantial amount of time to pass to be able to observe a violation. The code for these scenarios simply writes a message to standard output that time is passing, and we will see how to simulate fast-forwarding of time for validation purposes in later chapters.

In most cases, one would be interested in running all test cases, using the method: `static void runAllScenarios()`. The main class provided simply runs all scenarios using this method.

Exercise 3.4

Run the scenarios available in FiTS by calling `main` and go over the scenarios code to get an idea of their purpose.

chapter03

[8] In a way, these can be seen as test cases except that they do not include an oracle to check for correctness of the observed outcome.

3.6 Conclusions

FiTS is intended to provide a reasonably realistic financial transactions management system to illustrate how verification of complex properties can be carried out. The system has been kept reasonably small so as to ensure that not too much effort is required to understand its inner workings, thus enabling us to track verification instrumentation we will be performing.

Chapter 4
Manual Monitoring

The most basic means of monitoring and verifying a system for soundness at runtime is to check for a condition which should hold at a particular point in the system's lifetime and act accordingly if the check fails. For instance, a soundness condition for `FiTS` can be that accounts can never be in the red, and one way of checking this is to check the balance of an account every time it is changed. This is typically done through the use of assertions, which can be seen as the most basic of monitors. More complex properties which depend on the history of the system execution can also be handled in a similar manner.

In this chapter the reader is exposed to means of manual instrumentation of monitoring and verification, starting from simple assertions and progressing to more involved checks which require additional information to resolve, such as historic context, timing, or judgement on multiple objects at a time. Using `FiTS`, we show how assertions suffice for simple checks whereas for the more complex ones, all the necessary scaffolding scattered across the monitored software pushes the simple idea to its breaking point.

4.1 Monitoring using Assertions

Software developers are typically already familiar with assertions making them an ideal starting point for introducing monitoring. Referring back to the `FiTS` example, let us consider Property 3(the properties can be found on page 24): *"No account may end up with a negative balance after being accessed"*. Since the `balance` variable is marked as private, we only need to check this property in the methods of the `BankAccount` class. Furthermore, we can limit our checks to the methods `deposit` and `withdraw`, which are the only two methods which write to the variable. It suffices to add the assertion after

C. Colombo, G. J. Pace, *Runtime Verification*,
https://doi.org/10.1007/978-3-031-09268-8_4

each assignment to the balance variable (or just at the end of the methods if we do not require such a fine-grained check).[1]

Java already provides an assert command which takes a Boolean condition as parameter: assert balance >= 0. Java assertions can also take an additional string parameter to describe the error in case of a violation e.g. assert balance >= 0: "P3 violated". By default, the effect of assertions is ignored in Java, but they can be turned on to raise an exception when the condition does not hold (by using the Java flag -enableassertions or -ea). When using runtime monitoring on live systems, neither solution may be ideal. Not doing anything obviously defeats the purpose of using runtime verification in the first place, while interrupting execution and throwing a runtime verification exception would require the whole system to be re-engineered with this in mind, which is not always easy to do (this will be discussed further in Chapter 11). We provide an additional assertion function in a static Assertion class which simply reports the error string to standard error and allows execution to proceed e.g. Assertion.check(balance >= 0, "P3 violated"):

```
public class Assertion {
  public static void alert(String errorString) {
    System.out.println(errorString);
  }

  public static void check(Boolean condition, String errorString) {
    if (!condition) alert(errorString);
  }
}
```

Of course, if all monitor checks were this simple, this book would not exist. Let us consider Property 2:*"The transaction system must be initialised before any user logs in"*. This is a temporal property since it checks that events occur in a particular sequence. Most programming languages do not provide native support to reason about the history of execution or about the control flow of the program itself. The most straightforward way of implementing this property in Java would require us to use a Boolean variable to keep track of whether or not the transaction system has been initialised. This also means that the monitoring code of a single property is no longer present in a single location in the code. As a minimum, code for monitoring such a property would appear in four different locations:

1. A declaration of a Boolean variable which is visible to both the transaction system initialisation code and the login method.

[1] Ideally, the balance attribute is only set through a setter method, and we just check for correctness whenever it is updated. However, keep in mind that when adding property checks, we may not want to refactor the code-base, and checks would have to be added in a manner that matches the code at hand.

2. Initialisation of the variable to *false*. This might be combined with the previous point, however the monitor might need to be reset (e.g. in the case of testing) necessitating separate initialisation.
3. Update the Boolean variable once the transaction system is initialised.
4. Checking the status of the Boolean variable before accepting a user login.

Having monitoring logic spread across four parts of the code base is far from ideal. This is perhaps the biggest argument in favour of runtime verification: separation of concerns — keeping the monitoring and verification logic separate from the system code is usually a good idea. If they are not separated, the monitoring and verification logic remains so intertwined with the system logic that, for all intents and purposes, it becomes a part of the system rather than acting as a separate and external observer. Indeed, the extra code will end up increasing complexity of the system, possibly leading to additional programming errors.

As a first step towards separating monitoring and verification concerns, we will add the new logic to the `Verification` class where all the monitoring-related logic will be kept. We will still need some additional code spread calling the verification object strewn across the system, but they would simply involve calls to the `Verification` class methods rather than having the monitoring logic inlined in the system.[2]

The four monitoring logic components mentioned above are incorporated into the `Verification` class as follows:

1. We declare a static Boolean variable `fitsHasBeenInitialised` representing whether the initialisation method has already been called.
2. We provide a `setupVerification()` method which resets verification by resetting `fitsHasBeenInitialised` to false.
3. A method `fitsInitialisation()` is provided to be called from the initialisation method in `TransactionSystem`, and which sets the variable `fitsHasBeenInitialised` to true.
4. Finally, a method `fitsOpenSession()` to be called when a session is opened in `UserInfo` is executed. The method will assert that the variable `fitsHasBeenInitialised` is true in the FiTS login function.

[2] This separation of monitoring logic from the system is an important first step towards full separation of concerns. This simple property we are discussing already justifies its use, but it can be further justified in more complex scenarios where the system logic may use different programming languages, separate machines, or even be executed across different areas of a network. By allowing all such components to access the monitoring logic through an API, substantially simplifies the programming of monitors.

The code for the Verification class is thus as follows:

```
public class Verification {

  private static Boolean fitsHasBeenInitialised = false;

  // Called to start verification
  public static void setupVerification() {
    fitsHasBeenInitialised = false;
  }

  // Called from TransactionSystem.initialise
  public static void fitsInitialisation() {
    fitsHasBeenInitialised = true;
  }

  // Called from UserInfo.openSession
  public static void fitsOpenSession() {
    Assertion.check(fitsHasBeenInitialised, "P2 violated");
  }
}
```

In addition, the code of the system is changed to perform calls to the relevant verification methods as required as shown below:

```
public class BackEnd {

  public void initialise() {
    Verification.fitsInitialisation();
    ...
  }
  ...
}

public class UserInfo {

  public Integer openSession() {
    Verification.fitsOpenSession();
    ...
  }
  ...
}
```

Exercise 4.1

In the code provided, we have already included monitoring and verification for a number of properties. In this exercise we will examine this code and implement it for other properties:

1. Recall Properties 1, 3, and 8:
   - Property 1: *"Only users based in Argentina can be gold users"*.
   - Property 3: *"No account may end up with a negative balance after being accessed"*.
   - Property 8: *"The administrator must reconcile accounts every 1000 attempted outgoing external money transfers or an aggregate total of one million dollars in attempted outgoing external transfers (attempted transfers include transfers requested which never took place due to lack of funds)"*.

   Check the monitoring code provided for these properties.
2. Implement monitoring and verification code for Property 4: *"A bank account approved by the administrator may not have the same account number as any other bank account already existing in the system"*.
3. Recall Property 5 *"Once a user is disabled, he or she may not withdraw from an account until the administrator enables them again"*.Implement monitoring and verification code for this property using the user mode as kept track of by the system.
   However, we may choose not to trust this information because, after all, there may be a bug precisely in the code managing the user modes. Change your monitoring and verification code to keep track of the user mode within the verification class independently of the main system.

chapter04

The last exercise, in which we got to choose whether to trust the data kept track of by the system, or whether it made more sense to implement our own state tracking is a common dilemma when using runtime verification. The former is much more efficient and allows us to monitor and verify with much lower overheads but at the cost of not being able to identify bugs in the part of the main system which keeps track of such state and other parts that depend on it. Conversely, the latter is much safer and allows us to discover such bugs, but will have the state of the monitoring component grow as the number of users grows. Which is the way to go? In practice, it depends on how much monitoring overheads one can afford to have, and how one chooses to manage the risks involved — balancing the trust one has in the parts of the system which we are not checking against their criticality.

4.2 Parameterised Properties

If we choose to keep track of the state in the monitoring component, with a property such as: *"Once a user is disabled, he or she may not withdraw from an account until the administrator enables them again"*(Property 5), the monitoring state may no longer be of fixed size (such as a Boolean or integer variable). With this example, one would have to keep track of each user, thus resulting in monitoring overhead which grows as the number of users increases.

One way of looking at such properties is that we are effectively monitoring each user independently of each other, replicating the property for every user. Such properties are typically referred to in the literature as *parameterised properties*. In object-oriented programming, such properties would typically be specified on a per-object basis.

A naïve way of monitoring parameterised properties is by adding one or more attributes to the `Verification` class. For instance, you may have solved the last exercise by keeping track of the set of users which are disabled, updating the set whenever a user is enabled or disabled and using that information whenever a user tries to withdraw funds. Another way is to recognise that the monitoring is being done per-user, and to keep a monitoring state for each user. This is a more general solution which is preferable.

One can implement this additional state per-object by piggybacking on the existing object structure instances of which we are monitoring. For example, for Property 5, we can add a Boolean attribute to the `UserInfo` class definition which is manipulated and checked from the `Verification` object. The `UserInfo` class would be changed as shown below:

```
public class UserInfo {

  // Additional monitoring state
  public Boolean MONITORING_isEnabled = false;

  public void makeDisabled() {
    Verification.userMakeDisabled(this);
    ...
  }

  public void makeEnabled() {
    Verification.userMakeEnabled(this);
    ...
  }

  public void withdrawFrom(String accnum, Double amount) {
    Verification.userWithdrawal(this);
    ...
  }
}
```

Note that the Verification object methods will now receive a reference to the UserInfo object in order to be able to check and manipulate the new monitoring state variable. The corresponding Verification class would then be implemented in the following manner:

```
public class Verification {

  public static void userMakeDisabled(UserInfo u) {
    u.MONITORING_isEnabled = false;
  }

  public static void userMakeEnabled(UserInfo u) {
    u.MONITORING_isEnabled = true;
  }

  public static void userWithdrawal(UserInfo u) {
    Assertion.check(u.MONITORING_isEnabled(), "P5 violated");
  }
}
```

Exercise 4.2

Use this approach to monitor Properties 6 and 9:

- Property 6: *"Once greylisted, a user must perform at least three incoming transfers before being whitelisted".*
- Property 9: *"A user may not have more than 3 active sessions at any point in time".*

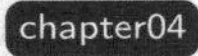

While adding new attributes for monitoring purposes may be convenient, it is in conflict with the rule-of-thumb of keeping monitoring code separate from system code. The fact that the additional data is stored in the system object under observation creates dependencies which we may prefer to do without. Furthermore, implementing the resetting of monitoring would require additional machinery to keep track of all the monitored UserInfo objects and iterate through them, resetting them individually.

The issue is further compounded by the fact that parameterised properties are very common in real-life scenarios and several of the other FiTS properties are, in fact, parameterised. A better way of handling this replicated monitoring state is to keep a mapping from the objects replicated (UserInfo in this case) and the monitoring state in the Verification class, for instance, using a HashMap structure:

```
public class Verification {

  private static HashMap<UserInfo, Boolean> userEnabled =
    new HashMap<UserInfo, Boolean>();

  public static void setupVerification() {
    userEnabled = new HashMap<UserInfo, Boolean>();
  }

  public static void userMakeDisabled(UserInfo u) {
    userEnabled.put(u, false);
  }

  public static void userMakeEnabled(UserInfo u) {
    userEnabled.put(u, true);
  }

  public static void userWithdrawal(UserInfo u) {
    Assertion.check(
      userEnabled.getOrDefault(u, false),
      "P5 violated"
    );
  }
}
```

One may argue that this is nothing more than an implementation of a set to represent the collection of enabled users. However, as the replicated state increases in complexity, this approach reaps benefits. One can even go one step further and refactor the `Verification` class to represent explicitly the fact that it is monitoring and verifying multiple users with the same property, and implement a per-user verification solution:

```
public class VerificationUserInfo {
  private Boolean isEnabled = false;

  public void userMakeDisabled() {
   isEnabled = false;
  }

  public void userMakeEnabled() {
   isEnabled = true;
  }

  public void userWithdrawal() {
    Assertion.check(isEnabled, "P5 violated");
  }
}
```

This verification unit is, in turn, replicated within the main `Verification` class:

```
public class Verification {

  private static HashMap<UserInfo, VerificationUserInfo> userVerifier =
    new HashMap<UserInfo, VerificationUserInfo>();

  public static void setupVerification() {
    userVerifier = new HashMap<UserInfo, VerificationUserInfo>();
  }

  public static void userMakeDisabled(UserInfo u) {
    userVerifier.computeIfAbsent(
      u, k -> new VerificationUserInfo()
    ).userMakeDisabled();
  }

  public static void userMakeEnabled(UserInfo u) {
    userVerifier.computeIfAbsent(
      u, k -> new VerificationUserInfo()
    ).userMakeEnabled();
  }

  public static void userWithdrawal(UserInfo u) {
    userVerifier.computeIfAbsent(
      u, k -> new VerificationUserInfo()
    ).userWithdrawal();
  }
}
```

Note the use of `computeIfAbsent` which will add an entry to the hash table if the `UserInfo` instance does not yet have an entry associated with it. The monitoring and verification is delegated to the instance for that user. Note that resetting of the whole verification state is now straightforward to achieve — by iterating through the monitored users in the hash table resetting them individually.

This approach may seem overkill if one is monitoring a property as simple as this one, but it starts to pay dividends as the monitoring state and logic becomes more complex. The advantages become even more noticeable if we are monitoring multiple properties per object e.g. if, in conjunction with Property 5 (*"Once a user is disabled, he or she may not withdraw from an account until the administrator enables them again"*), we were to monitor Property 9 (*"A user may not have more than 3 active sessions at any point in time"*), both of which check properties per instance of `UserInfo`.

Exercise 4.3

Implement monitoring and verification of the following properties using this last approach:

- Property 6: *"Once greylisted, a user must perform at least three incoming transfers before being whitelisted"*.
- Property 7: *"No user may request more than 10 new accounts in a single session"*.
- Property 9: *"A user may not have more than 3 active sessions at any point in time"*.
- Property 10: *"Logging can only be made to an active session (i.e. between a login and a logout)"*.

Keep in mind that some of the properties will require a replicated verification object other than `UserInfo`. For example, Properties 7 and 10 would best be handled with a verification object replicated per `UserSession`.

chapter04

When using this method, it is important to keep in mind how structures such as `HashMap` internally work in Java. For instance, `HashMap` internally uses the `equals` method to compare keys[3]. The default `equals` method for `UserInfo` compares pointers to the structures and would work as expected. However, had we overridden the `equals` method we could have had clashes arising between different objects which is not desirable. This is not an issue which arises in FiTS, but it is worth keeping in mind when one implements monitors in such a manner for more complex systems.

As a final note, it is worth observing that the per-object approach is very useful in an object-oriented language such as Java. In other paradigms which do not use the notion of objects, one may have to look at alternative ways of parameterising properties — for instance, one may replicate monitors on a per-process basis in Erlang, or on a per-function call invocation in C.

4.3 Conclusions

The idea of programming monitors directly as part of the main system code-base is a convenient way of implementing runtime verification. However, even when monitoring basic properties, the simple nature of assertions quickly results in complex arrangements and with code scattered throughout the code-base. Adding the properties directly into the monitored system makes

[3] In the process, it also uses the method `hashCode` but ultimately relies on the `equals` method to distinguish between objects having the same hash code.

it difficult to separate system from the monitoring and verification code, meaning that any changes in the properties automatically involve changes in the underlying system. This arrangement contributes to the complexity of the logic which the monitor is attempting to verify, possibly making it more bug-prone rather than error-free. In the next chapter we look at how aspect-oriented programming can help in solving this problem.

Chapter 5
Aspect-Oriented Programming

In the previous chapter we have been manually adding monitoring and verification code into the system code. For instance, when we wanted to ensure that FiTS has been initialised before allowing users to log in, a new Boolean flag was created, updated whenever relevant events were observed, and checked with every login attempt.

While this approach works, the system and the monitoring logic become intertwined. Although we can move verification logic away from the system and store it in a separate verification data structure, we would still have to add monitoring code in the system to notify the verification code whenever significant events happen. Sometimes we would prefer to do away even with such manual inlining.

This problem is not new or unique to runtime verification; it is the problem of *separation of concerns* — the principle that design should attempt to split a computer program into parts such that different modules address different concerns. Whilst some concerns can be separated relatively easily e.g. the user interface from the backend code, other concerns are by their very nature more *cross-cutting*. A frequently-used example of such a concern is that of logging software events, which *cross-cuts* across various modules in the system.

The challenge of introducing runtime verification to a system shares much with the challenge of adding logging system-wide, except that additional work must also be done to check for adherence to properties.

Whilst software engineering practices help to alleviate the problems of cross-cutting concerns, *aspect-oriented programming* (AOP) is a technique proposed specifically to address this challenge. In essence, AOP allows the programming of concerns separate from those of the main system, and weaving them together automatically based on code pattern matching.

As we will see in this chapter, this will allow us to specify monitoring and verification without any changes to the main system code, thus enabling us to achieve a more complete separation of concerns than that achieved until now.

C. Colombo, G. J. Pace, *Runtime Verification*,
https://doi.org/10.1007/978-3-031-09268-8_5

5.1 The Basics of Aspect-Oriented Programming

In this section, we will introduce the basic notions underlying AOP. Throughout, we will be giving concrete examples using AspectJ, an open source AOP tool for Java programs.

5.1.1 Joinpoints and Pointcuts

To enable automated code weaving, AOP requires the programmer to identify a number of points during the execution of a system where new code is to be injected. The supported types of such points, called *joinpoints*, vary depending on the programming language (and the programming paradigm of the language) used in the system at hand, but common points include the moment of entry into and exit from method calls, exception throws, variable assignments, class initialisation, constructor execution, etc.

Consider, for example, joinpoints encountered in an execution of the scenarios in FiTS. Even if we limit ourselves to method entry and exit points, we would get a huge number of joinpoints traversed:[1]

```
entry: void Main.main(String[] args)
entry: void Scenarios.runAllScenarios()
...
entry: void FrontEnd.ADMIN_createUser(String name, String country)
entry: Integer TransactionSystem.addUser(String name, String country)
exit: Integer TransactionSystem.addUser(String name, String country)
entry: UserInfo TransactionSystem.getUserInfo(Integer uid)
exit: UserInfo TransactionSystem.getUserInfo(Integer uid)
entry: void UserInfo.makeDisabled()
exit: void UserInfo.makeDisabled()
exit: void FrontEnd.ADMIN_createUser(String name, String country)
...
exit: void Scenarios.runAllScenarios()
exit: void Main.main(String[] args)
```

The information is voluminous, and once one adds extra information about these joinpoints, such as the target object (the object associated with the method), parameter values, and return value, it becomes crucial that we ought to be very selective as to which joinpoints to capture.

AOP employs *pointcuts*, essentially joinpoint filters expressed using pattern matching. For example, the pointcut to match just before calls to the

[1] The whole list of method entry and exit points is almost 35 million lines long. Even if one were to limit oneself to entry and exit points to methods in our code (i.e. not calls to methods from libraries), it would still be over 11 million lines long.

`makeDisabled()` method from any class and with any return type can be written as follows:

```
before (): call(* *.makeDisabled())
```

AspectJ allows a granular way of identifying entry and exit joinpoints. The `before` keyword indicates that we want to capture an entry point to a method (the corresponding keyword for exit points is `after`), while the `call` keyword indicates that the joinpoint refers to where (and when) the signature is called (as opposed to where and when it starts executing, in which case one would use the `execution` keyword). Note the use of the asterisk symbol to match any class and return type. Some AOP tools, including AspectJ, also allow for partial identifier names, e.g. `*.*User*()`, to match any joinpoint whose parameterless method call includes the substring `User`. If we want to modify this pointcut to capture the exit from any method call no matter how many parameters it takes, we would write it as follows:

```
after (): call(* *.*User*(..))
```

5.1.2 Advice and Code Injection

The utility of AOP is that the programmer may specify code, known as *advice*, to be executed at a set of identified joinpoints. For instance, if used for logging information, one could add advice to log information with the pointcut in which we are interested:

```
before (): call(* *.initialise(..)) {
  System.out.println("Entry to initialise");
}
```

Similarly, for monitoring purposes, we can use entry joinpoints to execute code to invoke the verification logic. Note that a joinpoint may match multiple pointcuts, thus triggering the execution of more than one advice (typically in the order in which they appear in the AOP source).[2]

[2] AspectJ also has a notion of precedence, which defines which joinpoints take precedence over others when they trigger simultaneously. The precedence rules, albeit being simple, may give rise to unresolvable orders which could stop parts of an aspect from firing. This topic is beyond the scope of this book, and we simply note that in order to avoid such problems from arising, it suffices to separate `before` from `after` pointcuts rather than interleave them together.

5.1.3 Adding Attributes and Methods

In addition to the ability to inject code at runtime, AOP can also be used to inject new attributes to existing classes. For instance, an AOP instruction such as: `public Boolean UserInfo.enabled = false;` would add a Boolean variable called `enabled` to the `UserInfo` class. Similarly, one would write `private Boolean *.isBeingLogged = false;` to insert a Boolean variable `isBeingLogged` to all classes.

Furthermore, one can add new methods to classes such as the following:

```
public void *.toggleLogging() {
isBeingLogged = !isBeingLogged;
}
```

This adds a method to toggle the `isBeingLogged` variable we added earlier. One can combine the code from these examples to enable the logging of calls to initialise methods, but also allowing for the logging to be turned on and off:

```
private Boolean *.isBeingLogged = false;

public void *.toggleLogging() {
isBeingLogged = !isBeingLogged;
}

before (): call(* *.initialise(..)) {
  if (isBeingLogged) System.out.println("Entry to initialise");
}
```

A collection of such directives — pointcuts, advice together with attribute and method additions — is called an *aspect*, hence the name *aspect-oriented programming*. This brief introduction to AOP should suffice to allow us to use it for monitoring and verification purposes in the rest of the chapter.

5.2 Using AspectJ for Monitoring

Now that we have introduced AOP concepts, we can turn our attention to their application to runtime monitoring and verification. Let us start by looking at Property 2:*"The transaction system must be initialised before any user logs in"*, showing how it can be runtime verified using AOP.

We start off with the `Verification` class we used earlier:

```
public class Verification {

  private static Boolean fitsHasBeenInitialised = false;

  // Called to start verification
  public static void setupVerification() {
    fitsHasBeenInitialised = false;
  }

  // Called from BackEnd.initialise
  public static void fitsInitialisation() {
    fitsHasBeenInitialised = true;
  }

  // Called from UserInfo.openSession
  public static void fitsOpenSession() {
    Assertion.check(fitsHasBeenInitialised, "Property 2 violated");
  }
}
```

Now, in order to invoke `Verification` class methods, instead of adding code to the main system, we can write AspectJ aspects:

```
public aspect Property2 {

  before (): call(* TransactionSystem.setup()) {
    Verification.fitsInitialisation();
  }

  before (): call(* UserInfo.openSession()) {
    Verification.fitsOpenSession();
  }
}
```

It is worth noting an AspectJ aspect is enclosed in an `aspect` block. Note that the system code has remained untouched and that we are injecting code using a separate script specifying new code through the use of aspects.

Exercise 5.1

The code presented in this section to handle Property 2 is included in the repository, including the verification and AspectJ code (in the `Properties.aj` file — as in the case of Java, the filename must match the aspect name). Make sure you have AspectJ installed on your machine, after which you can compile the Java sources with the AspectJ aspects using your favourite IDE (with appropriate Java and AspectJ plugins if required). Execute and make sure that the code works as you would expect.

chapter05

The previous example was particularly simple because the advice did not have to make any reference to the object captured by the joinpoint. Now consider if we were to use AspectJ to add monitoring code for the Property 3:*"No account may end up with a negative balance after being accessed"*. There are various ways in which we can capture points-of-interest in the code from which to trigger a check for the `balance` variable.

If we add joinpoints to method calls, we may want to trigger a check for a negative balance at the end of every call the `withdraw` and `deposit` methods, since these are the only places where the `balance` variable is updated.

```
after ():
  call (* BankAccount.withdraw(..)) || call (* BankAccount.deposit(..)) {
   Verification.balanceChanged(?.getBalance());
}
```

At this point note that instead of writing two rules (as we could have done), we are using the AspectJ disjunction operator `||` to take the union of two pointcuts. However, note that we lack a reference to the `BankAccount` whose `deposit` or `withdraw` method is called.[3] This object can be captured using `target` keyword as shown below:

```
after (BankAccount a): target(a) &&
   (call(* BankAccount.withdraw(..)) ||
    call(* BankAccount.deposit(..))
   ) {
      Verification.balanceChanged(a.getBalance());
}
```

First of all, note that we are using the conjunction operator `&&` which takes the intersection of two pointcuts. Next note that the pointcut `target(a)` captures the target object to variable `a`, whose type is specified in the parentheses before the colon.

It is worth adding that we are using the getter function `getBalance` which provides access to `balance`. If you want to access `balance` itself (perhaps you might worry that there may be a bug in the `getBalance` if it were more complex) you could check it directly. However, since `balance` is a private field, the aspect will not have access to it. In order to allow the aspect to access private attributes, we can change the aspect's visibility from `public` to `privileged`:

[3] It is worth mentioning that if you were to use `this` in an advice, it would capture the aspect, and not the class matching the pointcut.

```
privileged aspect Property3 {

  after (BankAccount a): target(a) &&
    (call(* BankAccount.withdraw(..)) ||
     call(* BankAccount.deposit(..))
    ) {
    Verification.balanceChanged(a.balance);
  }

  ...
}
```

So far, we have been checking the value of `balance` at the end of methods which change its value. However, we could have also chosen to capture whenever an assignment is made to the variable. This can be done by capturing `set` as a pointcut.[4]

```
after (BankAccount a): target(a) && set(* BankAccount.balance) {
  Verification.balanceChanged(a.balance);
}
```

Yet one other way in which this property can be monitored is to check that withdraw is never called with an amount which exceeds the remaining balance. This can be monitored using AspectJ as follows:

```
before (BankAccount a, Double amount):
  target(a) && call(* BankAccount.withdraw(..) && args(amount) {
    Verification.balanceChanged(a);
}
```

In this case, we are using `args` to bind the parameters to names to be used within the advice. Note that the types are identified within parentheses in the initial part of the pointcut. Also, if the caught method takes more than one argument, they may all be named in the `args` parameter e.g. `args(a,b,*)`, using the asterisk is used to ignore irrelevant ones.

Exercise 5.2

Implement the different approaches discussed above to runtime checking the property Property 3: *"No account may end up with a negative balance after being accessed"*. Modify the aspects so as to use the `Verification` class used in the previous chapter for this property.

[4] Make sure that the aspect is privileged since `balance` is private.

Using the `Verification` classes you created for the previous chapters, adapt your solutions using AspectJ to inject monitoring code for the following properties:

- Property 1: *"Only users based in Argentina can be gold users".*
- Property 4. *"A bank account approved by the administrator may not have the same account number as any other bank account already existing in the system".*
- Property 5 *"Once a user is disabled, he or she may not withdraw from an account until the administrator enables them again".*
- Property 6. *"Once greylisted, a user must perform at least three incoming transfers before being whitelisted".*
- Property 7. *"No user may request more than 10 new accounts in a single session".*
- Property 8. *"The administrator must reconcile accounts every 1000 attempted outgoing external money transfers or an aggregate total of one million dollars in attempted outgoing external transfers (attempted transfers include transfers requested which never took place due to lack of funds)".*
- Property 9. *"A user may not have more than 3 active sessions at any point in time".*
- Property 10. *"Logging can only be made to an active session (i.e. between a login and a logout)".*

Combine your properties so that they are all checked as the system executes.

chapter05

In this chapter, we have continued with the code organisational principle which we presented in the previous chapter, separating the verification code into an independent module and having the main code simply call relevant methods notifying the verification engine of relevant events and data. The down side of this is that we may have to replicate structures inherent in the system in order to store monitoring state e.g. the use of hash tables to keep track of the different user accounts.

While this organisational principle was crucial when we manually instrumented monitors and verifiers, this is less so with the use of AOP. Since the new code is still organised separately from the main system through the use of aspects, we have the luxury of being able to choose whether we inject it directly as part of the system code or — as we have done till now in this chapter — inject only calls to the separate verification module.

Consider the instrumentation of monitoring and verification code for Property 5: *"Once a user is disabled, he or she may not withdraw from an account until the administrator enables them again".* To keep the verification code separate, we would have to keep track of the mode of each user through the use of a set (tracking which users are enabled) or a hash table (mapping users to their mode). In contrast, if we were to inject code directly into the system, we can do away with this structure by using that inherent in the system. By adding a new attribute to `UserInfo` to keep track whether the user is enabled, we can check the property from within that class with no further data structures:

```
// This adds a Boolean attribute to class UserInfo
private Boolean UserInfo.isEnabled = false;

before (UserInfo u): call(* UserInfo.makeDisabled(..)) && target(u) {
  u.isEnabled = false;
}

before (FrontEnd fe, Integer uid):
  call(* FrontEnd.ADMIN_enableUser(..)) && target(fe) && args(uid) {
    fe.getBackEnd().getUserInfo(uid).isEnabled = true;
}

before (UserInfo u): call(* UserInfo.withdrawFrom(..)) && target(u) {
  Assertion.check(u.isEnabled, "P5 violated");
}
```

Still, the advantage of piggybacking on the system objects without the need of creating additional data structures is usually outweighed by the fact that it makes it impossible to execute the monitor independently from the system, e.g. for offline monitoring (see Chapter 12) or when one is monitoring properties across distributed systems (see Chapter 13). Furthermore, note that once the object is destroyed, the monitor is automatically destroyed with it, which is not always desirable.

Exercise 5.3

Implement monitors using this approach to runtime check the following properties:

- Property 6: *"Once greylisted, a user must perform at least three incoming transfers before being whitelisted"*.
- Property 9: *"A user may not have more than 3 active sessions at any point in time"*.
- Property 10: *"Logging can only be made to an active session (i.e. between a login and a logout)"*.

chapter05-inlined

5.3 Advanced AOP Considerations

There are various other AOP features that can be useful for runtime monitoring and verification. In this section we discuss a number of these in order to point the reader in the right direction when applying runtime verification to more complex systems. The material mentioned in this section is not required to follow the rest of the book.

Call vs. execution pointcuts: We have already briefly mentioned the difference between `call` and `execution` pointcuts, which match on the caller or on the callee side respectively. This means, for instance, that `execution` cannot be used on third party libraries, while `call` does not match on the `main` method. An important difference is that the `call` joinpoint has information about both the caller's and the callee's context while the `execution` joinpoint only has access to the callee's.

Advanced pointcut bindings: We have already seen how to bind the target object on which an intercepted method has been called and extract the arguments passed — information which is crucial for runtime verification. However, there are also other elements which may be useful for more sophisticated runtime verification: (i) accessing the return value of the method: `after () returning (Type value)`; (ii) triggering upon an exception being thrown and accessing the exception: `after () throwing (Exception e)`; (iii) binding the thread object on which the method is running: `thread(t)`; and (iv) the object bound by `this` keyword at the point of matching: `this(o)`.

Advanced pointcut conditions: An important consideration in pointcut specification is that pointcuts are matched at runtime, meaning that one can specify conditions to distinguish between different runtime contexts. Two useful examples in the context of runtime verification are the following:

- To avoid runtime monitors capturing events generated by their own advice (and potentially resulting in livelock), we can add the condition `!cflow(adviceexecution())` to the pointcut in order to ensure that no joinpoints resulting from the control flow of the advice (i.e the monitoring code in our case) will be matched.
- Another useful case is when one wants to capture the initial call to a recursive method but not the subsequent recursive calls. This can be implemented using the `cflowbelow` condition for a recursive method `f` for instance as follows: `call(* *.f(..)) && !cflowbelow(call(* *.f(..))` indicating that any pointcuts originating from the control flow below the recursive method will be ignored.

Both these examples use variations of `cflow`, which allows the specification of the control-flow context in which a pointcut will match.

Weaving modes: Once aspects have been specified, there are three points during which these can be weaved into the target system:

Compile-time: The aspect code is weaved into the system at compile-time, resulting in binaries already including the weaved-in aspects. This is the default way weaving works in AspectJ.

Post-compile-time: If the source code is not available, for example in the case of third party libraries, aspects can be weaved directly into the bytecode.

Load-time: The third option is to postpone weaving up till the point when the JVM loads a class. The advantage of this approach is the added flexibility of leaving the bytecode untouched and deciding how to weave the target system through a configuration file. On the other hand, there is naturally an efficiency penalty every time a class is loaded.

In the rest of the book we will be using the default compile-time weaving approach in order to have binaries already including the aspects.

Exercise 5.4

Experiment with some of these AOP features on FiTS.

chapter05-inlined

5.4 Conclusions

AOP provides a way of programming cross-cutting concerns in a modular fashion, and is perfectly suited to support runtime verification, with monitoring permeating the whole system. In reality, the AOP features required for runtime monitoring are rather basic but, as shown in the previous section, its use can become rather intricate. Keeping in mind that AOP can modify and control all system behaviour, judicious use of AOP is crucial.

Up till this point in the book, we have seen how one can manually add coded runtime monitors and verifiers. Although AOP proves to be an excellent tool to separate such code from that of the main system, we still have to intricately program verification code in a manual manner.

The upcoming chapters aim at making the building of runtime monitors and verifiers safer by allowing the user to use progressively higher-level specification languages which would be automatically compiled to aspects and verification code. This helps in two ways. Firstly, by hiding the AOP details from the user, thus eliminating inadvertent errors in the aspects being programmed. Secondly, the high level of the specification language will allow us to express properties more abstractly — in a way, closer to the English language descriptions, and have them compiled into Java verification code automatically. In a sense, this allows us to take the separation of concerns principle even further by having an implementation-agnostic property language independent of the system logic.

Chapter 6
Event Guarded Command Language

6.1 Introduction

We have already seen how the use of aspect-oriented programming can help in separating concerns between system functionality and monitoring logic. However, we are still programming specification logic by hand and this can be a source of error as we encode more complex properties. One way of addressing this is to introduce higher-level means of writing specifications.

In this chapter, we look at a way in which properties can be expressed in a more understandable manner, moving away from low-level use of AOP and allowing for a rule-based approach to specification writing. We will build on this in later chapters to provide even more abstract means of writing specifications.

6.2 Event Guarded Command Language: The Syntax

Once we have identified which events of interest may arise during the system's execution, we can express verification logic as a set of instructions telling the verification engine what to do when a particular event is observed at runtime. The runtime verification language we will present facilitates expressing such instructions. The *Event Guarded Command Language* (EGCL)[1] we will present and use in this chapter allows us to express verification logic in terms of *rules* of the following form:

```
event | condition -> { action }
```

[1] The name and syntax of the language are inspired by Edsger Dijkstra's *Guarded Command Language*, which used guarded commands similar to the ones we use.

C. Colombo, G. J. Pace, *Runtime Verification*,
https://doi.org/10.1007/978-3-031-09268-8_6

Each such rule provides the instructions (action) to perform whenever the rule's event is observed at runtime and the condition holds. To avoid defining a new language for the specification of conditions and actions, we use Java as the host-language to express the rules.

In terms of events, we will allow referring to the entry and exit point of a method using the `before method(parameters)` or `after method(parameters)` notation. The event will trigger the moment control is about to be diverted to the named method or just after returning from the named method respectively.[2] Consider an EGCL rule which triggers whenever a withdrawal exceeding \$1000 is made:

```
before *.withdraw(Double amount) | amount > 1000
  -> { Assertion.alert("High withdrawal attempt: "+amount+" dollars"); }
```

Note that the event component of the rule uses syntax similar to AspectJ. In particular, an asterisk (`*`) is used as a wildcard to match any class or method. Identified parameters are given a type and name as in normal method declarations, and can be referred to in the condition and action code. If the parameters will not be used, we simply use double period (`..`).

Furthermore, a rule may also identify the *target* of the event by giving it a name in the event part of the rule. Consider a rule to identify when `withdraw` is invoked with a value larger than the balance of the account:

```
before *.withdraw(Double amount) target (UserAccount a)
  |  a.getBalance() < amount
  -> { Assertion.alert("Attempt to withdraw more than funds held"); }
```

It is rarely the case that a single rule suffices to specify a property. Typically we need multiple rules to be written, each firing when the relevant event is detected. If an event matches more than one rule, all the rules are triggered and fire in the order in which they appear in the list. Consider the following two EGCL rules used to capture violations of Property 2:*"The transaction system must be initialised before any user logs in"*, segregating some of the verification logic into the `Verification` class:

```
after BackEnd.initialise() | true
  -> { Verification.fitsInitialisation(); }
before UserInfo.openSession() | !Verification.isInitialised()
  -> { Assertion.alert("P2 violated"); }
```

[2] Needless to say, the approach we will be exploring in this chapter is not bound to the choice of these events, and one can add other events of interest e.g. an exception being thrown, so long as they can be captured by the monitor when they happen.

As we have done in the previous examples we keep the verification logic in the static `Verification` class keeping track whether or not FiTS has been initialised:

```
public class Verification {

  private static Boolean fitsHasBeenInitialised = false;

  // Called to start verification
  public static void setupVerification() {
    fitsHasBeenInitialised = false;
  }

  // Called from BackEnd.initialise
  public static void fitsInitialisation() {
    fitsHasBeenInitialised = true;
  }

  // Check whether initialised
  public static Boolean isInitialised() {
    return fitsHasBeenInitialised;
  }

}
```

Note that it is not always straightforward whether we should capture the event before or after the method call. In the example above, we assume that the system may call these methods concurrently, and it is thus important that the initialisation code has terminated (thus using `after`) before a session can be opened (using `before`). Later on in the book (see Chapter 11), we will discuss the importance of when to capture violation events depending on whether we want to intervene (before it happens) or simply report the violation (after it happens).

Note that the first rule used in the specification triggers whenever the `initialise` method is called with the condition effectively being `true`. In such cases, we may leave out the condition altogether as shown below:

```
after UserInfo.initialise() | -> { Verification.fitsInitialisation(); }
```

Exercise 6.1

Express the following FiTS properties using EGCL rules:

- Property 1: *"Only users based in Argentina can be gold users"*.

- Property 2: *"The transaction system must be initialised before any user logs in".*
- Property 3: *"No account may end up with a negative balance after being accessed".*
- Property 4: *"A bank account approved by the administrator may not have the same account number as any other bank account already existing in the system".*
- Property 8: *"The administrator must reconcile accounts every 1000 attempted outgoing external money transfers or an aggregate total of one million dollars in attempted outgoing external transfers (attempted transfers include transfers requested which never took place due to lack of funds)".*

chapter06

6.3 Compiling into AspectJ

Now that we have introduced the syntax for the rules, we can start considering how to instrument them into the system. One way of doing this is by generating AspectJ code. We will combine the rules and the support code of the monitoring logic in a single script, from which our tool will generate the AspectJ and Java code separately.

In order to achieve this, apart from the rules shown in the previous section, an EGCL script will include two additional parts: (i) the verification code which will be used to support the verification logic in the rules; and (ii) the imports and declarations necessary for the AspectJ code to work (e.g. the package name declaration and the list of imports). An EGCL script is structured into these three parts using the `VERIFICATIONCODE`, `PRELUDE` and `RULES` section headers:

```
VERIFICATIONCODE
// Verification logic

PRELUDE
package rv;
import fits.*;
// Any other prelude code

RULES
// Rules go here
```

We can combine the rules and verification code we saw in the previous chapter to runtime verify Property 2 (*"The transaction system must be initialised before any user logs in"*) into the following script:

```
VERIFICATIONCODE

public class Verification {

  private static Boolean fitsHasBeenInitialised = false;

  // Called to start verification
  public static void setupVerification() {
    fitsHasBeenInitialised = false;
  }

  // Called from BackEnd.initialise
  public static void fitsInitialisation() {
    fitsHasBeenInitialised = true;
  }

  // Check whether initialised
  public static Boolean isInitialised() {
    return fitsHasBeenInitialised;
  }

}

PRELUDE

package rv;
import fits.*;

RULES

after BackEnd.initialise() | -> { Verification.fitsInitialisation(); }
before UserInfo.openSession() | !Verification.isInitialised()
  -> { Assertion.alert("P2 violated"); }
```

Exercise 6.2

Write the rules from the previous exercise as a EGCL script using the syntax shown above.

chapter06

We will now turn our focus to building a tool to transform such scripts into executable AspectJ and Java code. The process flow of the tool is shown in Figure 6.1. Given the system (FiTS) and the specification (the EGCL script) the tool will create two new files — the AspectJ code (*Properties.aj*) and the verification code (*Verifier.java*). When these files are combined with the original FiTS, we obtain a monitored version of the system.

We now turn our attention to the writing of the compilation tool, which takes a script and produces the AspectJ and Java verification code. In order

Fig. 6.1 Runtime verification tool-flow in the case of EGCL

to focus on the runtime verification aspect, we provide basic code for parsing and manipulating EGCL scripts in chapter06 under the `MyRVTool` folder. This code is organised as follows:

- `Main.java`: A simple class which reads and parses an EGCL script and outputs it again to standard output.
- `EGCLScript.java`: A class to represent, manipulate and handle EGCL scripts. The constructor `EGCLScript(String filename)` parses an EGCL script from a file possibly throwing an exception in case of a syntax error. The class has getter methods to access the different parts of the script: (i) the `VERIFICATIONCODE` section can be accessed as a string using `String getVerificationCode()`; (ii) the `PRELUDE` part can be accessed as a string using `String getPrelude()`; and (iii) the `RULES` can be accessed as a list of `Rule` objects (see below) using `ArrayList<Rule> getRules()`. Note that the first two return strings since we do not intend to perform any further manipulation when processing them.
  Finally, the class has a `String toString()` method which returns the rules in the EGCL script corresponding to the parser-generated data structure.
- `Rule.java`: This class is used to represent, access and manipulate an EGCL rule.
  The getters for the condition and action components of an EGCL rule are: `String getCondition()`, and `String getAction()`. The rule event is accessed using the `Event getEvent()` method.
  The class also contains a `String toString()` method to output the rule using the EGCL syntax we are using.
- `Event.java`: This class is used to represent events as they appear in the rules.
  The name of the event (everything apart from the parameters and the target information) can be obtained using the `String getEvent()`. Information about the parameters can be obtained using `ArrayList<String> getParameterVariables()` and `ArrayList<String> getParameterTypes()`, and information about the target is obtained through the two methods: `String getTargetVariable()` and `String getTargetType()` (both returning null if the rule does not specify a target). Finally, to check whether the

event has a `before` or `after` modality, we provide two methods: `Boolean isBefore()` and `Boolean isAfter()`.
The method `String toString()` is also provided.

- `AspectJScript.java`: This class somewhat simplifies the generation of AspectJ files, by ensuring that rules are written in an order which does not cause AspectJ precedence issues. It provides the method `void addAfterAJRule(String rule)` adds the given `after` rule to the aspect, and similarly `void addBeforeAJRule(String rule)` adds a `before` rule. To add any additional code within the aspect body (before the rules), the method `void addToAspectPrelude(String code)` is provided, and `void addToFilePrelude(String code)` adds the code *before* the code of the aspect (e.g. to add imports at the start of the AspectJ file). Finally, a `toString()` method is provided to create a string with the content of an AspectJ file with all the code and rules given.
- `RuleParser.java`: This class contains the logic to parse the rules. We need not be concerned with the content of this class for the purposes of this book.

Exercise 6.3

Review the code provided and then:

1. Implement a `String toAspectJ()` method to the `EGCLScript` class. **Hint:** Create an `AspectJScript` object to add rules and code, and pass this object to a corresponding `toAspectJ(AspectJScript ajs)` method in the `Rule` class, which will update the object accordingly. Then use the `toString` method in the `AspectJScript` class to generate the code. Parts of this are already provided in the accompanying code.
2. Modify `Main.java` to read a script file and output the AspectJ file and the verification class Java file.
3. Test the tool by compiling the properties from the previous exercise.
4. Finally, test the output of your compiler to monitor the scenarios as before.

chapter06/MyRVTool

6.4 Parameterised Properties in EGCL

In the previous two chapters we considered the monitoring of parameterised properties which apply to multiple entities, such as Property 6: *"Once greylisted, a user must perform at least three incoming transfers before being whitelisted"*. With such properties, one has to keep track of the state of the property for multiple objects. For instance, in the case of this property, one has to track whether each user has been disabled in order to be able to conclude whether a withdrawal is legitimate.

Using EGCL, every time we have such a property, one has to add additional code to handle the replicated monitor. We can monitor Property 6 using the following script managing the monitoring state explicitly:

```
after UserInfo.makeGreylisted(..) target (UserInfo u)
  |  !Verification.isGreylisted
  -> {
    Verification.isGreylisted = true;
    Verification.countTransfers.put(u,0);
  }

before UserInfo.depositTo(..) target (UserInfo u) |
  -> {
    if (Verification.isGreylisted)
      Verification.countTransfers.put(
        u, Verification.countTransfers.get(u)+1
      );
  }

before UserInfo.makeWhitelisted(..) target (UserInfo u)
  |  Verification.isGreylisted
       && Verification.countTransfers.get(u) < 3
  -> { Assertion.alert("P6 violated"); }

after UserInfo.makeWhitelisted(..) target (UserInfo u) |
  -> { Verification.isGreylisted = false; }
```

The `Verification` class would then keep a mapping from users to a Boolean value (tracking whether or not a user is greylisted) and another one to an integer (tracking the number of transfers performed by each user since being greylisted) to keep track of the monitoring state for each user.[3] However, as you may have noticed when implementing monitors for parameterised properties, such additional code follows a pattern for different properties. In view of this, we can choose to allow scripts to specify that a collection of rules should be applied per-object and have the compiler provide the additional code to handle it.

For a given property, what we need to know is (i) the type of the object we are parameterising over; (ii) the monitoring memory which is to be replicated for each of these objects; and (iii) the default value which the memory starts

[3] You may question whether it would make more sense to use the information from the `UserInfo` class to know whether or not a user is greylisted. We choose to manage that information as part of the monitor just in case there is a bug in the code handling that variable.

off in whenever an event on a new target is observed. We can incorporate such information as part of the EGCL scripts to enable us to compile parameterised properties automatically using three keywords `foreach`, `keep`, and `defaultTo`. The script we just saw can be expressed as follows using this new notation:

```
foreach target (UserInfo u)
  keep (Boolean greylisted defaultTo false,
        Integer countTransfers defaultTo 0
  ) {

  after UserInfo.makeGreylisted(..)
    |  !greylisted
    -> { greylisted = true;
         countTransfers = 0; }

  before UserInfo.depositTo(..)
    |  greylisted
    -> { countTransfers = countTransfers + 1; }

  before UserInfo.makeWhitelisted(..)
    |  greylisted && countTransfers < 3
    -> { Assertion.alert("P6 violated"); }

  after UserInfo.makeWhitelisted(..) |
    -> { greylisted = false; }
}
```

It is worth noting that we assume that for a given instance of the property, the observed events will target that same object. In fact, the rules inside a `foreach` statement do not have their own individual target, but rather inherit the target from the `foreach` they appear in (`UserInfo` in this case). Secondly, note that more than one variable can be used for monitoring memory, and this is expressed as a tuple after the `keep` keyword. For simplicity, we will only allow `foreach` statements at the top level, and we do not allow the operator to be nested.[4]

Transforming this code into actual monitoring code now involves adding hash tables as required, despite the fact that the specification is free from any such references, thus allowing us to focus on the logic of the property itself.

[4] In practice, nesting `foreach` statements can be useful, for example to parameterise over every account of every user.

Exercise 6.4

Write scripts for each of the following properties using EGCL, making use of the parameterisation features:

- Property 5: *"Once a user is disabled, he or she may not withdraw from an account until the administrator enables them again".*
- Property 9: *"A user may not have more than 3 active sessions at any point in time".*
- Property 10: *"Logging can only be made to an active session (i.e. between a login and a logout)".*

chapter06-parameterised

6.5 Compiling Parameterised EGCL

We will now look at how such scripts can be compiled. Let us see what needs to be done with the EGCL script for Property 6. Firstly, data structures are to be created to store information about each instance of the property using a hash table from the target specified in the `foreach` statement to the type of the memory state. We do so for each `keep` variable, also defining initialisation methods for the hash tables:

```
static HashMap<UserInfo, Boolean> greylisted =
  new HashMap<UserInfo, Boolean>();

static void initialisegreylisted() {
  greylisted = new HashMap<UserInfo, Boolean>();
}

static HashMap<UserInfo, Integer> countTransfers =
  new HashMap<UserInfo, Integer>();

static void initialisecountTransfers() {
  countTransfers = new HashMap<UserInfo, Integer>();
}
```

This code can be part of the AspectJ code (just after the contents of the `PRELUDE` declaration) to allow us to access the information directly from the pointcuts. We now turn our sights on the actual rules. Consider the second rule in the script:

```
before UserInfo.depositTo(..)
  | greylisted
  -> { countTransfers = countTransfers + 1; }
```

The event can be readily handled, using the target from the encapsulating `foreach` to obtain the following pointcut:

```
before (UserInfo u): call(* UserInfo.depositTo(..)) && target(u) {
  ...
}
```

Finally, we must implement the condition and action parts of the rule. We face a problem with references in the code (the condition and action) to the monitoring state variables in the rules — the code must be modified so as to access the values through the hash tables in the AspectJ version. Assignments to these variables must be replaced by calls to the hash table `put` method, and any reading of the variables must be done via a call to the hash table `get` or even better the `computeIfAbsent`[5] method. Transforming the condition into an `if` statement, we obtain:

```
before (UserInfo u): call(* UserInfo.depositTo(..)) && target(u) {
  if (greylisted.computeIfAbsent(u, v -> false))
    countTransfers.put(u,countTransfers.computeIfAbsent(u, v -> 0) + 1);
}
```

Using this pattern, we can transform all the rules to obtain the following pointcuts:

[5] This method also takes a function as input which assigns a value to the key when absent, e.g. `greylisted.computeIfAbsent(u, v -> false)` signifying that if user object u is not found in the hash table, the default value `false` is associated with the key and inserted in the table (as though `greylisted.put(u, false)` was called), and then returns the value — either the retrieved one, or the default value if the key was not present in the hash table.

```
after (UserInfo u): call(* UserInfo.makeGreylisted(..)) && target(u) {
  if (!greylisted.computeIfAbsent(u, v -> false)) {
    greylisted.put(u,true);
    countTransfers.put(u,0);
  }
}

before (UserInfo u): call(* UserInfo.depositTo(..)) && target(u) {
  if (greylisted.computeIfAbsent(u, v -> false))
   countTransfers.put(u,countTransfers.computeIfAbsent(u, v -> 0) + 1);
}

before (UserInfo u):
  call(* UserInfo.makeWhitelisted(..)) && target(u) {
  if (greylisted.computeIfAbsent(u, v -> false) &&
       countTransfers.computeIfAbsent(u, v -> 0) < 3) {
    Assertion.alert("P6 violated");
  }
}

after (UserInfo u): call(* UserInfo.makeWhitelisted(..)) && target(u) {
  greylisted.put(u, false);
}
```

You will find an enhanced version of the EGCL code we saw earlier in chapter06-parameterised under the `MyRVTool` folder. The important changes from the code we saw earlier are the following:

1. `EGCLScript.java`: In addition to the data and methods discussed earlier, a list of replicated rulesets is also stored. This list can be accessed using the method `ArrayList<Foreach> getForeaches()`.[6]
2. `Foreach.java`: This class handles the parameterised rulesets. Methods are provided to obtain information about the `foreach` statement: (i) the target's identifier and type using `String getTargetVariable()` and `String getTargetType()`; (ii) the replicated state variable names using `ArrayList<String> getReplicatedStateVars()`, the types of the variables `ArrayList<String> getReplicatedStateTypes()` and default values `ArrayList<String> getReplicatedStateDefaults()`; and (iii) the rules appearing in the `foreach`, using `ArrayList<Rule> getRules()`.
3. `Rule.java`: This class now has code to allow for renaming of variables in conditions and actions. The method `replaceReplicatedStateAssignment` takes (i) the code to transform (as a string); (ii) the variable name used for the `foreach` target; (iii) a list of state variable names; and (iv) their default values. The method returns a string with the code modified such

[6] We separate the simple rules from the replicated rulesets for simplicity. Note that this changes the order of the rules, and if the order in which rules are to be executed matters, we may be changing the behaviour. In the rest of the book we will assume that rules are written such that rule reordering does not change behaviour.

that assignments to the monitoring state are replaced by a call to the `put` method on the relevant hash table entry. Similarly, the method `replaceReplicatedStateReference` takes the same parameters but replaces read instances of that variable with calls to `computeIfAbsent` as discussed earlier.[7]

Exercise 6.5

Review the new code provided, and extend as follows:

1. Implement a `void toAspectJ(AspectJScript)` method to handle replicated state in the `Foreach` class and modify the `toAspectJ` methods of the `Rule` and `EGCLScript` classes if and as required. Remember to add the declarations to the top of the AspectJ generated. Note that rules are now to be compiled differently depending on whether they are at the top level or within a `foreach`.
2. Using the previously modified `Main.java`, test the tool by compiling the properties from the previous exercise.
3. Finally, test the output of your compiler to monitor the scenarios as before.

chapter06-parameterised/MyRVTool

6.6 Conclusions

EGCL has provided us with a simple abstraction to move away from the nitty-gritty detail of aspect-oriented programming allowing us to present the monitoring logic in a structured, rule-based approach. Furthermore, by adding parameterisation of rules at the EGCL level, we also allow for such properties to be expressed in a simpler manner.

However, one can argue that the abstraction provided by EGCL is still limited. In the next chapter, we will move on to support properties expressed as automata, providing a graphical representation as well as a further level of abstraction of the monitoring state.

---

[7] Note that the methods provided are kept simple and only variable changes through an assignment statement are identified. They will not work if a more sophisticated data structure is updated through method calls.

Chapter 7
Symbolic Automata

In our quest to raise the abstraction level of specifications and keeping them separate from the system implementation, we have introduced EGCL — allowing the writing of specifications and properties without having to instrument them as aspects.

Whilst this simplifies property writing considerably, the state of the monitors has to be manually expressed (in the form of Java variables and updates thereof), which increases the complexity of reading and understanding specifications. This is especially true when these state checks and updates appear in different rules.

Recall Property 2:*"The transaction system must be initialised before any user logs in"*. We expressed this property using EGCL as follows;

```
after BackEnd.initialise()
   -> { Verification.fitsInitialisation(); }
before UserInfo.openSession() | !Verification.isInitialised()
   -> { Assertion.alert("P2 violated"); }
```

In order to understand this specification, one has to look within the `Verification` object to find a Boolean variable being declared, checked and maintained throughout its interface.

In this chapter we will look at ways of encoding parts of the monitoring state as part of the specification language without having to program them explicitly in Java. We will use a visual notation, expressing properties as *finite state automata* which, as we will see, allow us to encode progress in the property graphically in a manner which can be automatically compiled to generate Java code handling the state.

C. Colombo, G. J. Pace, *Runtime Verification*,
https://doi.org/10.1007/978-3-031-09268-8_7

7.1 Automata for Monitoring

Finite state automata are frequently used to model and specify computation. In our case, we will use automata to describe in a visual manner how the system's state evolves. Consider the automaton shown in Figure 7.1, which tracks whether a particular user session is open or closed. The automaton has two states (NoSession and ActiveSession), and starts off in the NoSession state (note the incoming arrow annotated with *start*). Observing a call to `openSession()` when in the NoSession state triggers a transition (the arrow between states) and updates the current state to ActiveSession. Conversely, observing `closeSession()` when in ActiveSession changes the state back to NoSession. Any other events observed and not on any outgoing transition are simply ignored, leaving the automaton in the same state.[1]

We can generate Java code to keep track of the state from such a description, relieving us of the need to program it explicitly and thus simplifying our specifications.

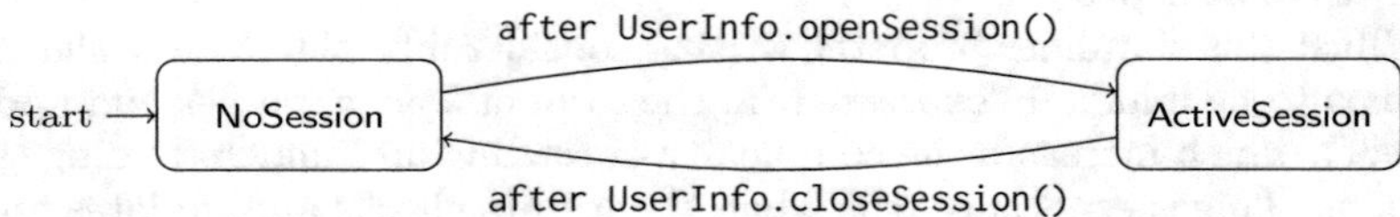

Fig. 7.1 Finite state automaton keeping track of a user session.

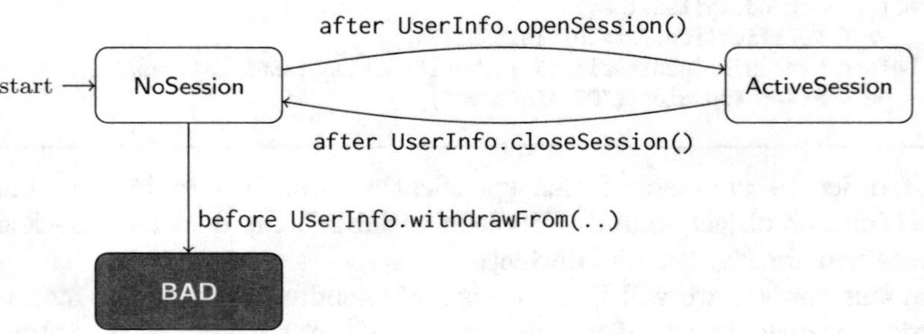

Fig. 7.2 Finite state automaton identifying `withdraw` attempts without an open session.

The automaton shown in Figure 7.1 is not a specification, since it does not distinguish between desirable and undesirable behaviour. We could change

[1] It is worth pointing out that this automaton takes into account just events coming from a single user who cannot open more than one session at a time, ignoring events from users. We will look into this deeper later on in this chapter.

the interpretation of the automaton such that events not appearing in outgoing transitions are deemed to be incorrect behaviour, but that would lead to intricate specifications, having to add transitions handling greylisting, whitelisting, creation of accounts and any other allowed behaviour. A more sensible approach, is that we mark undesirable behaviour explicitly by adding a *bad state* and modify the automaton so as to have unexpected behaviour send us to this state. This is the solution we will adopt, shading bad states to distinguish them from normal ones. We can express the property that a `withdraw()` command should only be called during open session as the automaton shown in Figure 7.2.

Let us now turn to Property 2:*"The transaction system must be initialised before any user logs in"*. Recall that using EGCL we explicitly kept track whether or not the transaction system was initialised. We can express this property using finite state automata, without the need for such additional code, as shown in Figure 7.3.

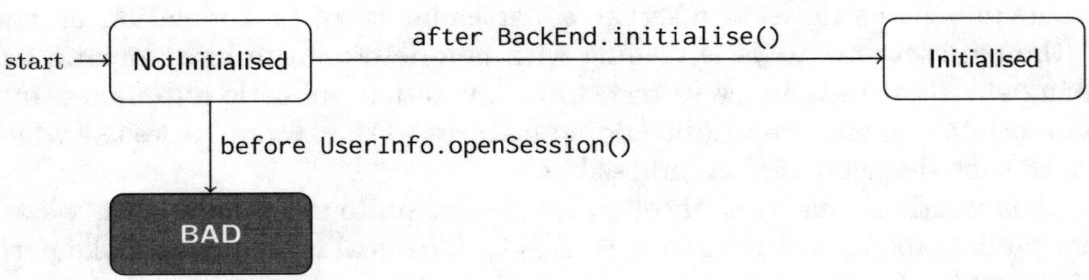

Fig. 7.3 Finite state automaton of Property 2.

We will express automata specifications using a textual notation. Below is the script corresponding to the automaton expressing Property 2:

```
VERIFICATIONCODE

PRELUDE
package rv;
import fits.*;

AUTOMATA
property starting NotInitialised {
  NotInitialised >>>
    after BackEnd.initialise()
      >>> Initialised
  NotInitialised >>>
    before UserInfo.openSession()
      >>> BAD["P2 violated"]
}
```

The structure of the script is similar to that of EGCL scripts: (i) a `VERIFICATIONCODE` section with any verification code as may be required; (ii) a `PRELUDE` section with code to be included in the AspectJ generated; and (iii) an `AUTOMATA` section with a description of automata expressing the properties we want to monitor.

Each automaton is described in a `property` block which identifies the name of the initial state after the `starting` keyword. Transitions are expressed using the >>> ternary operator, with `A >>> e >>> B` denoting that there is a transition from state `A` to state `B` when event `e` is observed. States do not have to be explicitly declared, but are simply referred to in the transitions. Finally, a special state `BAD` is used to capture violations, with a string parameter passed in square brackets describing the violation. A script can have multiple property blocks, with one automaton per block.

We will assume that the automaton is deterministic — given a current state and event, there is never more than one transition that can trigger. Non-deterministic automata are not ideal for monitoring, since we would like an unambiguous decision whether a particular trace is a violation or not. Although there are ways of dealing with non-determinism in automata, e.g. using standard techniques to transform a non-deterministic automaton into a deterministic one, we will be adding features to the automata we use which make such determinisation impossible.

It is worth noting that through the use of finite state automata, we are not adding any expressive power to EGCL. The goal is simply to make parts of the state of a property implicit i.e. without having to program that aspect of the monitoring state explicitly.

Exercise 7.1

Modify `FiTS` to have a `BackEnd.shutdown()` method, after which the transaction system should no longer allow any sessions from opening. Modify Property 2 to allow `openSession()` after initialisation, but before a shutdown. Express the modified property as a finite state automaton and modify the scenarios to include a trace that satisfies this extended property, and one that violates it.

7.2 Symbolic Automata

Finite state automata simplify the writing of specifications since part of the monitoring state is implicitly represented using the states of the automaton. If the property we are expressing covers a finite number of possible situations,

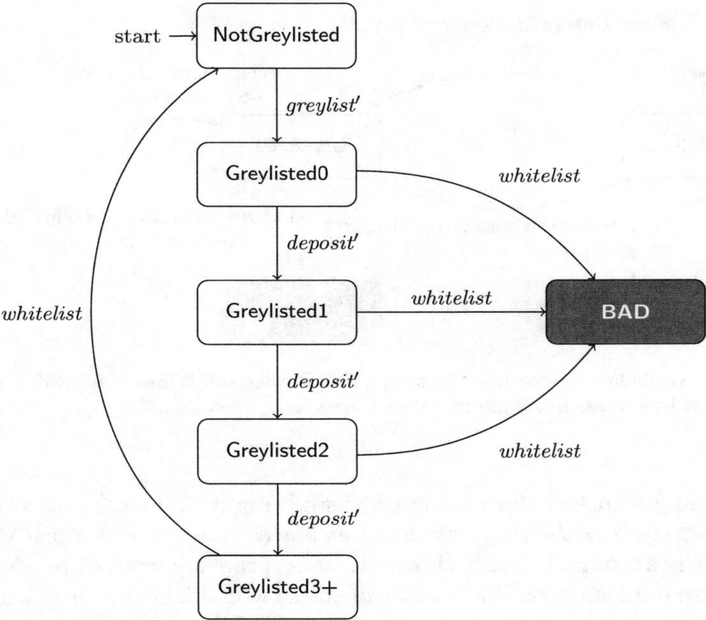

Fig. 7.4 Finite state automaton of Property 6 for the single user case. The tags on the transitions refer to the following events: (i) *greylist′* refers to `after UserInfo.makeGreylisted()`; (ii) *whitelist* refers to `before UserInfo.makeWhitelisted()`; and (ii) *deposit′* refers to `after UserAccount.deposit(..)`.

we can usually readily express it using an automaton. However, automata are not a universal solution for expressing specifications.

- Any monitoring state that can range over an unlimited number of possible values cannot be represented using the states of a finite state automaton. Consider a property which limits user activity based on the balance of transfers performed (total withdrawn less total deposited) in a single session e.g. they may not transfer to certain accounts if they have a transfer balance exceeding \$1000. Such a property must keep track of the balance of transfers performed by the user which is technically infinite and may thus not be directly represented using the states of the automaton.
- The number of states of the automaton must also be manageable in practice. Consider Property 6:*"Once greylisted, a user must perform at least three incoming transfers before being whitelisted"*. The property can keep track of the number of incoming transfers using the state of the automaton. We can keep separate states for: (i) when the user is not greylisted; (ii) the user has just been greylisted; (iii) the user has received one incoming transfer since being greylisted; (iv) the user has received two incoming transfers since being greylisted; (v) the user has received three or more

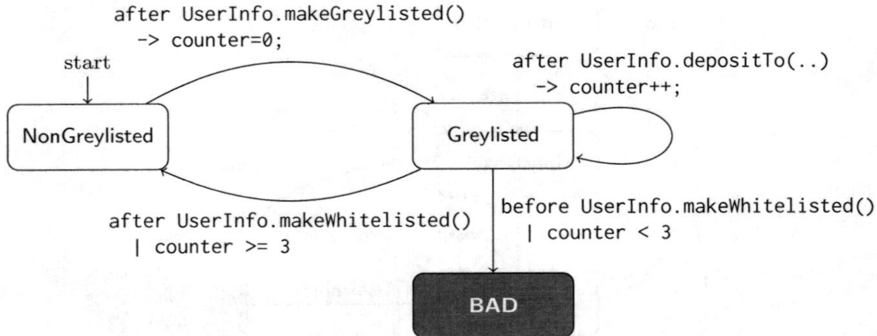

Fig. 7.5 Symbolic state automaton monitoring Property 6: *"Once greylisted, a user must perform at least three incoming transfers before being whitelisted".*

incoming transfers since being greylisted. Figure 7.4 shows an automaton to keep track of the property using six states (the five states just identified plus the extra bad state). However, if the property were to be changed to require ten transfers, the number of states would increase proportionately, making the automaton more complex to understand and more expensive to monitor.

In order to address parts of the monitoring state ranging over an infinite or a large domain, we will use symbolic automata which include (i) state encoded using the actual states in the automaton; and (ii) state encoded using variables whose value may be checked and updated by the automaton (hence the use of the term *symbolic*).

Consider Property 6:*"Once greylisted, a user must perform at least three incoming transfers before being whitelisted"*, running with a single user for simplicity. We can use states to identify when the user has been greylisted, but add an integer variable `count` to keep track of the number of transfers performed since their greylisting. The symbolic automaton handling Property 6 is shown in Figure 7.5. Some of the transitions check and update the value of the count, and are annotated with the event which triggers it, a condition which must be satisfied and an action to carry out — effectively an EGCL rule.[2]

When a user is greylisted, the state changes from NonGreylisted to Greylisted and the count of deposits is reset to zero. Once greylisted, deposits to the user's accounts will increment the count, and an attempt to whitelist the user

[2] The property is not clear what happens if the user is blacklisted while being greylisted. Does the property disallow blacklisting a user who has just been greylisted and then whitelisting them despite no transfers having been made? Clearly this is a matter of interpretation, but for the sake of this chapter, we will adopt the interpretation that once greylisted, a user must make three transfers before being whitelisted no matter if their status is changed in the meantime.

will trigger a transition either (i) going to a bad state if less than three deposits have been performed since greylisting; or (ii) succeeding and returning to the start state Whitelisted if three or more deposits were performed. If the user is blacklisted, the state changes back to NonGreylisted and the counter is reset.

In order to describe symbolic automata, we extend the syntax shown in the previous section to be able to handle rules on transitions and the declaration of symbolic state variables as shown below:

```
VERIFICATIONCODE

PRELUDE
package rv;
import fits.*;

Integer count;

AUTOMATA
property starting NonGreylisted {
  NonGreylisted >>>
    after UserInfo.makeGreylisted() -> count = 0;
      >>> Greylisted
  Greylisted >>>
    after UserInfo.depositTo(..) -> count++;
      >>> Greylisted
  Greylisted >>>
    after UserInfo.makeWhitelisted() | count >= 3
      >>> NonGreylisted
  Greylisted >>>
    before UserInfo.makeWhitelisted() | count < 3
      >>> BAD["P6 violated"]
}
```

Note that the symbolic monitoring state is declared in the AspectJ prelude, but could also have been declared as part of the verification code (as part of a `Verification` class) and accessed accordingly i.e. as `Verification.count`.

As in the case of simple finite state automata, we will require the transitions to be mutually exclusive in order to ensure that monitoring is deterministic. To understand better this requirement, consider for instance, if the condition of the `UserInfo.makeWhitelisted()` transition going from the Greylisted to the bad state were to be modified to `count < 4`, and the whitelisting event happens after exactly three deposits, the monitor can change non-deterministically to either the NonGreylisted state or the bad one. Earlier in this chapter we noted that non-deterministic finite state automata (with no symbolic state) can be transformed into an equivalent deterministic one. However, with symbolic automata and the use of actions on transitions, such a transformation becomes impossible to do in general. One could attempt to use backtracking to simulate the non-deterministic choice, but this would require taking snapshots of state variables to allow backtracking to their previous values, making the space cost excessively high.[3]

Exercise 7.2

Express Property 8 (*"The administrator must reconcile accounts every 1000 attempted outgoing external money transfers or an aggregate total of one million dollars in attempted outgoing external transfers (attempted transfers include transfers requested which never took place due to lack of funds)"*) in terms of a symbolic automaton using the notation explained above.

chapter07

7.3 Compiling Symbolic Automata

In order to monitor specifications expressed as symbolic automata, we will have to generate additional code directly from the properties. Ignoring the symbolic aspect (the extra monitoring variables, the conditions and actions on the transitions), we would still need to add additional information to keep track in which state the automaton resides as monitoring proceeds.

If we were to compile the automaton down to AspectJ code, we would need to create a new monitoring state variable to keep track of which state the automaton resides during execution. In addition, each transition of the form `Source >>> event >>> Destination` would result in (i) defining a pointcut which triggers when the `event` happens; (ii) we instrument code at that stage so that, if the current state is `Source`, then it is updated to `Destination`. For instance, consider the following transition:

```
NotInitialised >>>
  after BackEnd.initialise()
    >>> Initialised
```

This can be compiled to the following pointcut to instrument the code handling the change of state:[4]

[3] In addition, some actions cannot be undone or retried (along another non-deterministic path) — take, for instance, printing a message on the screen or sending an email.

[4] We could improve performance by enumerating the states in order to compare and assign integers, not strings. However, for simplicity, we will be using the strings throughout the chapter.

```
after (): call(* BackEnd.initialise()) {
  if (state.equals("NotInitialised"))
    state = "Initialised";
}
```

All the transitions would be dealt with in the same manner except for ones which go to the bad state, and which would not only update the state name, but also report the error string. Consider the following transition which goes to the bad state:

```
NotInitialised >>>
  before UserInfo.openSession()
    >>> BAD["P2 violated"]
```

This transition would be instrumented using the following pointcut:

```
before (): call(* UserInfo.openSession()) {
  if (state.equals("UnInitialised")) {
    state = "BAD";
    Assertion.alert("P2 violated");
  }
}
```

Instead of compiling straight to AspectJ code we can, however, compile to EGCL rules and use the compilation procedure we built in the previous chapter. In this manner the compilation of automata becomes simpler and cleaner. The two rules we presented above would thus compile to the following:

```
after BackEnd.initialise()
  |  state.equals("NotInitialised")
  -> { state = "Initialised"; }

before UserInfo.openSession()
  |  state.equals("UnInitialised")
  -> {
    state = "BAD";
    Assertion.alert("P2 violated");
  }
```

We can extend this compilation pattern to deal with symbolic automata in which transitions are decorated with EGCL rules. The use of EGCL as an intermediate compilation stage simplifies the translation of symbolic transitions: we simply (i) combine the transition condition with the condition checking the state of the automaton; and (ii) perform the action given on the transition rule with the code to update the state of the automaton. Consider the following transition:

```
Greylisted >>>
  after UserInfo.depositTo(..) | count < 3 -> count++;
    >>> Greylisted
```

This transition would be compiled to the following EGCL rule:

```
after UserInfo.depositTo(..)
  |  state.equals("Greylisted") && (count < 3)
  -> {
    count++;
    state = "Greylisted";
  }
```

Note the use of extra brackets around the condition appearing in the rule decorating the transition to avoid potential precedence problems. To show this translation on a complete example, the automaton shown in Figure 7.5 (and expressed in text format in Section 7.2), would be compiled to the following EGCL rules:

```
PRELUDE
. . .
Integer count;
String state = "NonGreylisted";

RULES
after UserInfo.makeGreylisted()
  |  state.equals("Whitelisted") && (true)
  -> {
    count = 0;
    state = "Greylisted";
  }

after UserInfo.depositTo(..)
  | state.equals("Greylisted") && (true)
  -> {
    count++;
    state = "Greylisted";
  }

after UserInfo.makeWhitelisted()
  |  state.equals("Greylisted") && (count >= 3)
  -> { state = "NonGreylisted"; }

before UserInfo.makeWhitelisted()
  |  state.equals(Greylisted) && (count < 3)
  -> {
    state = "BAD";
    Assertion.alert("P6 violated");
  }
```

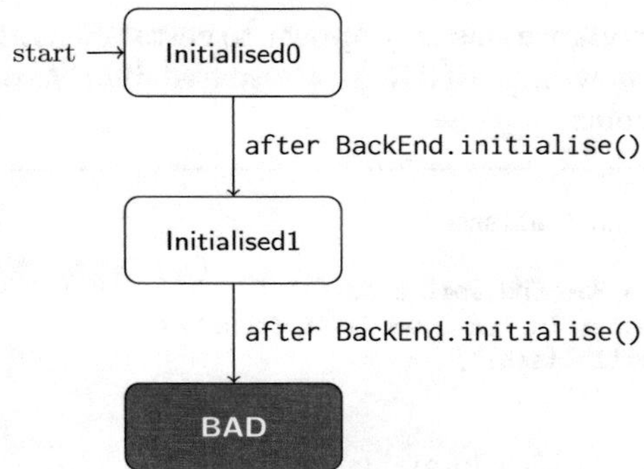

Fig. 7.6 Symbolic state automaton monitoring that FiTS should not be initialised more than once.

The state variables are declared at the start, together with the variable to store the current state of the automaton, which starts off assigned to the initial state.

Although this code handles the original automaton faithfully, there are cases where this compilation will not match the expected behaviour of the original automaton. Consider the symbolic automaton shown in Figure 7.6 which should trigger a violation if FiTS is initialised more than once. The automaton would be translated into the EGCL shown below:[5]

```
PRELUDE
...
String state = "Initialised0";

RULES
after BackEnd.initialise()
  | state.equals("Initialised0") && (true)
  -> { state = "Initialised1"; }

after BackEnd.initialise()
  | state.equals("Initialised1") && (true)
  -> {
    state = "BAD";
    Assertion.alert("Multiple initialisation");
  }
```

[5] We assume that the system has no variable named `state`.

Although at first glance this may appear to correctly model the automaton, a deeper look at how this EGCL is translated into AspectJ will uncover unexpected behaviour:

```
String state = "Initialised0";

after (): call(* BackEnd.initialise()) {
  if (state.equals("Initialised0") && (true))
    state = "Initialised1";
}

after (): call(* BackEnd.initialise()) {
  if (state.equals("Initialised1") && (true)) {
    state = "BAD";
    Assertion.alert("Multiple initialisation");
  }
}
```

Since the pointcuts are executed from top to bottom, a call to the method `BackEnd.initialise()` will trigger both rules in order: If it is the first time that `initialise()` is called, the condition of the first pointcut will be satisfied, updating the state to `Initialised1`. When the second pointcut is now matched, the condition is satisfied once again, resulting in a violation. The problem is that using this naïve translation, a single event can trigger a chain of transitions in the same automaton.

How can we solve this? One solution is to gather all transitions which trigger with the same event and code the rule resolution such that following a transition will automatically eliminate the rest e.g. using nested conditionals. Another solution, the one we will adopt, is to add an extra Boolean variable `hasTriggered` to keep track whether a transition has already triggered with the received event. Setting the variable to *false* before any event is processed, we can guard each rule with the condition that `hasTriggered` must be false, and set it to *true* once a rule triggers.

This latter solution can be incorporated into the EGCL by generating rules as follows:[6]

[6] We assume that the system has no variable named `hasTriggered`.

```
PRELUDE
...
String state = "Initialised0";
Boolean hasTriggered = false;

RULES
after BackEnd.initialise()
  |  !hasTriggered && state.equals("Initialised0") && (true)
  -> {
    state = "Initialised1";
    hasTriggered = true;
  }

after BackEnd.initialise()
  |  !hasTriggered && state.equals("Initialised1") && (true)
  -> {
    state = "BAD";
    Assertion.alert("Multiple initialisation");
    hasTriggered = true;
  }
```

The remaining question is how to reset the `hasTriggered` variable to `false` before any event is processed. We do so by initialising the variable to *false*, and resetting it back to *false* after processing *all* events, by adding the following two rules at the end of the generated EGCL script:

```
before *.*(..) | -> { hasTriggered = false; }
after  *.*(..) | -> { hasTriggered = false; }
```

This ensures that no more than one transition per automaton is triggered per event observed.

Let us now turn our focus to the compilation of symbolic automata. Starting from a script with a specification consisting of a number of automata, we can produce EGCL from which we can then produce the AspectJ (using the code we developed in the last chapter) to instrument the monitors. We provide basic code for parsing automata scripts in chapter07 under the `MyRVTool` folder. The code is organised as follows:

- `Main.java`: This class uses the code in the files described below, to read an automata script and pretty prints it back to standard output.
- `AutomataScript.java`: A class to represent, manipulate and handle automata scripts. The constructor `AutomataScript(String filename)` parses an automata script from the file with the given filename, throwing an exception if a syntax error is encountered.

  The class provides getter methods to access the different sections of the script: (i) `String getVerificationCode()` returns the `VERIFICATIONCODE` section as a string; (ii) `String getPrelude()` returns the `PRELUDE` part, also

as a string; and (iii) `ArrayList<Automaton> getAutomata()` returns a list of automata as given in the `AUTOMATA` section of the script. The structure of the `Automaton` class is given below.
A method called `String toString()` is also provided, returning the automata script as a string.

- `Automaton.java`: Each automaton is represented as an instance of this class.
  The class provides a getter for the start state: `String getStartState()`, and one for the transitions: `ArrayList<Transition> getTransitions()`. A method `String toString()` is also provided to output the automaton using automata syntax.
- `Transition.java`: This class is used to represent a transition. The source and destination state of the transition can be accessed using `String getSourceState()` and `String getDestinationState()` respectively, and the EGCL rule can be accessed using `Rule getRule()` method (see the documentation for the `Rule` class in Section 6.3 on page 58). A `String toString()` method is also provided.
- `AutomatonParser.java`: This contains code to parse automata, which we need not be concerned with.

Exercise 7.3

Review the code provided and then:

1. Assuming that the automata script has only one automaton, implement a `toEGCL()` method in the `Transition`, `Automaton` and `AutomataScript` classes. **Hints:** (i) Look at the constructors of the `EGCLScript` class to see how to create the object; (ii) Ensure that you add the EGCL rules to avoid multiple transitions triggering; and (iii) The declaration for the state variable can be appended to the `PRELUDE` code of the `EGCLScript` object.
2. Modify your code to handle the possibility of multiple automata being specified in the script. **Hint:** You will need a separate variable for each automaton to store their current states. Also, you will need a separate `hasTriggered` variable for each automaton to ensure that an event may trigger transitions in multiple automata, but not multiple ones in the same automaton. Make sure to avoid name clashes by using a unique element (e.g. a counter or the object memory address) in your variable names.
3. Test the tool by compiling the properties from the previous exercises in this chapter.
4. Finally, test the output of your compiler to monitor FiTS.
5. As an advanced exercise, change the state representation of the automata to use integers rather than strings. All states must first

be associated with an integer and state comparison and updates should use the integer representation instead of strings, thus improving performance.

chapter07/MyRVTool

7.4 Parameterised Automata

As we noted in the EGCL chapter, many properties are not about the system as a whole, but rather about every object of a particular type. In fact, even some of the examples we showed in this chapter were artificial in that we had to make assumptions limiting the system to a single user or a single session. For instance, Figure 7.5 showed an automaton which monitored Property 6:*"Once greylisted, a user must perform at least three incoming transfers before being whitelisted"*. However, as we noted, the monitor only works if we are monitoring a single user. Otherwise, deposits of other users would count in favour of the greylisted user, allowing them to be whitelisted if there were three or more deposits system-wide, and not just by the greylisted user as originally intended. The monitor of the property should, in fact, be replicated for every user.

In order to handle such per-entity properties in EGCL, we introduced a `foreach` quantifier together with a `keep` state variables declaration, which allow us to monitor properties for each instance of a class. In this section, we will extend automata to allow them to be replicated in the same manner as EGCL rules.

Back to the greylisting example, what we would like to have is an instance of the automaton shown in Figure 7.5 for each distinct instance of `UserInfo`. Each automaton should keep track of its own copy of `count`, using which it keeps track of the number of times the `depositTo` method is called for that particular user. We adopt syntax almost identical to that used in EGCL to express such replicated properties:

```
property foreach target (UserInfo u)
    starting NonGreylisted
    keep (Integer count defaultTo 0) {

  NonGreylisted >>>
    before UserInfo.makeGreylisted(..) target (UserInfo u) |
      -> count = 0;
  >>> Greylisted

  Greylisted >>>
    after UserInfo.depositTo(..) target (UserInfo u) |
      -> count = count + 1;
      >>> Greylisted

  Greylisted >>>
    before UserInfo.makeWhitelisted(..) target (UserInfo u)
      | count >= 3
      >>> NonGreylisted

  Greylisted >>>
    before UserInfo.makeWhitelisted(..) target (UserInfo u)
      | count < 3
      >>> Bad["P6 violated"]
}
```

This specification monitors each individual instance of `UserInfo` using the given automaton. Note that all the transitions in the replicated automaton must trigger on events arising from the target class, in this case `UserInfo`. Furthermore, a copy of each variable declared in the `keep` part of the definition, `count` in this case, is also kept for every instance.

Exercise 7.4

Write parameterised automata scripts for the following properties:

- Property 5: *"Once a user is disabled, he or she may not withdraw from an account until the administrator enables them again"*.
- Property 9: *"A user may not have more than 3 active sessions at any point in time"*.
- Property 10: *"Logging can only be made to an active session (i.e. between a login and a logout)"*.

chapter07-parameterised

7.5 Compiling Parameterised Automata

When an automaton appears replicated under a `foreach` construct, we must keep track of the `state` variable for every instance of that class. We can do this by using parameterisation in EGCL. The EGCL used to monitor Property 6 would be the following:

```
property foreach target (UserInfo u)
  keep (
    Integer count defaultTo 0,
    String state defaultsTo "Whitelisted"
  ) {

  after UserInfo.makeGreylisted()
    |  state.equals("Whitelisted") && (true)
    -> {
      count = 0;
      state = "Greylisted";
    }

  ...

  before UserInfo.makeWhitelisted()
    |  state.equals(Greylisted) && (count < 3)
    -> {
      state = "BAD";
      Assertion.alert("P6 violated");
    }
}
```

Note that (i) the translation from the automaton to EGCL remains unchanged; (ii) we add the `foreach` quantifier and `keep` variables from the automata to the parameterised EGCL; but (iii) we add a new replicated variable to the `keep` declaration to keep track of the state of the automaton.

The instrumentation of parameterised symbolic automata will use the same strategy as before — we start from a script describing a specification expressed as a number of replicated automata, from which we produce parameterised EGCL, which can then be compiled to AspectJ for instrumentation. The code for parsing and representing parameterised automata can be found in `chapter07-parameterised` under the `MyRVTool` folder. The code is organised identically to that described in Section 7.3 with the following differences:

- `AutomataScript.java`: Over and above what we had before, this class now provides an additional method returning the parameterised automata `ArrayList<ParameterisedAutomaton> getParameterisedAutomata()`. Automata which are not parameterised are still accessed using the method `ArrayList<Automaton> getAutomata()`.

- `ParameterisedAutomaton.java`: Parameterised automata are represented as instances of this class, with getters provided for the different components: (i) `String getTargetVariable()` and `String getTargetType()` return the name and type of the `target` variable; (ii) the names of the variables declared using `keep` can be accessed using `ArrayList<String> getReplicatedStateVars()`, their type using method `ArrayList<String> getReplicatedStateTypes()` and their default values using method `String getReplicatedStateDefaults()`; and (iii) `Automaton getAutomaton()` returns the structure of the underlying automaton.

Exercise 7.5

Review the code provided and then:

1. Using the approach you took in your solution to Exercise 7.3, change the code so as to handle parameterised automata. You may assume that there will be no name clashes with the variable you use to keep track of the automata states.
2. Ensure that `Main.java` reads an automaton script file (which may have parameterised automata) and outputs an AspectJ file and the verification Java class file.
3. Test the tool by compiling the properties from the previous exercise.

chapter07-parameterised/MyRVTool

7.6 Conclusions

Automata not only offer a visual means of expressing a property, but by doing so, they allow implicit encoding of a part of the property state thus simplifying it. For properties which have a notion of sequentiality i.e. in what order events are expected to happen, this abstraction is highly beneficial. On the other hand, when used to express complex properties, automata tend to quickly become difficult to understand.

In practice, it is impossible to define the perfect property specification language. It depends on the type of properties one needs to specify, the type of system being verified, the verification technique to be used, and many other variables. It is, however, important to be aware of the various options and their characteristics to use the right formalisation in the right context.

In the next two chapters, we present two formula-based logics which can be used to express certain properties succinctly and in a structured manner. However, as we will see, this comes at the price of a more involved compilation and monitoring process.

Chapter 8
Regular Expressions

Richer property specification languages allow us to express properties more succinctly, thus leaving less room for error. The complexity lies in building runtime monitors and verifiers for such languages. Visual specification languages, such as the automata we have seen in the previous chapter, can be useful to express many properties, but suffer from lack of *compositionality*: automata do not lend themselves well to being broken down into smaller parts which can be understood independently, and which can then be glued together to obtain more complex properties. Furthermore, automata do not allow us to build *structured properties*. The rise of structured programming was largely in response to undisciplined control flow in programs (particularly due to the use of the `GOTO` statement), and addressed such unnecessary complexity by enforcing structure — for example ensuring that overlapping loops are not possible unless strictly nested in each other. Transitions in automata are nothing but `GOTO` statements, and complex properties easily end up a mess of spaghetti joining different states.

Given that EGCL and symbolic automata lack such structure, the question is what property specification formalisms we can use which have these desirable features. By using a *language* in which to express properties, one which has both syntax and semantics which are structured,[1] we can achieve this goal.

One structured formalism which programmers are typically familiar with is that of *regular expressions*. In this chapter we look at how regular expressions can be used to express runtime verification properties, and how runtime monitors and verifiers can automatically be synthesised from such specifications.

[1] Structured syntax ensures that properties can be decomposed syntactically. By also ensuring that the semantics of the language is context-free and structured, we know that the meaning of an expression is not impacted by another independent expression. We lack such structure in both automata and EGCL specifications.

C. Colombo, G. J. Pace, *Runtime Verification*,
https://doi.org/10.1007/978-3-031-09268-8_8

8.1 Regular Expressions

Most programmers are familiar with regular expressions to describing patterns against which to match a string. In a way, they are effectively acting as string specifications based on the individual characters, and they efficiently classify strings into ones which match and ones which do not. If we see a string as a stream of characters, this is analogous to what our runtime verifiers do — they consume a stream of events, deciding whether or not the string of events seen until now matches the specification. In this section we will define regular expressions over events (rather than characters), in order to enable us to express system properties.

For those unfamiliar with regular expressions, we start by presenting their syntax and the semantics, after which we will show some examples of how they can be used as specifications. The syntax we will use to express regular expressions matches that which we will use in our tool. It consists of the following basic elements and operators:

Single event matching: The basic building block for monitoring regular expressions is the expression to match a particular named event happening once. We express this as the name of the event in square brackets e.g. `[after TransitionSystem.initialise()]` matching at the end of the initialisation method, and `[before UserInfo.depositTo(..)]` matching just before a deposit. For simplicity we will limit monitoring regular expressions to events (i) that are class-specific i.e. we will not have events which use `*` for the class name, such as `*.depositTo(..)`; and (ii) will not use methods which are overloaded with a different number of parameters or parameter types.[2]

Another way of matching a single event is to use *complemented events* which match any event other than a particular one. For example, we might want to know when any event other than the start of a call to `depositTo` method occurs, in which case we will write: `[!before UserInfo.depositTo(..)]`.

Finally, we will allow for the matching of *any* single event using the regular expression `any`. This and event complementation may not appear to be very useful in practice, but as we will see later on, these will be essential to express many properties.[3]

Choice: Given two regular expressions e and f, we can express a regular expression which matches when either of the two subexpressions matches

[2] An explanation as to why we make these limitations will be given later on in this chapter. Also, it is important to note that these are not limitations of the regular expression approach, but rather constraints that will make our implementation more straightforward.

[3] You may also be questioning why we are just using events with no conditions or actions in the events. After all, both in EGCL and automata, we found them to be helpful, if not essential. We will discuss why actions can cause problems later on in the chapter once we have have a better understanding of how regular expressions can be used and monitored.

as e + f. For instance, if we want to match when either a deposit or a withdrawal has occurred, we can write:

```
[before UserInfo.depositTo(..)] + [before UserInfo.withdrawFrom(..)]
```

Sequential composition: Another way of composing regular expressions is to match them in sequence. Given two regular expressions e and f, the regular expression e ; f matches a sequence of events if it can be split into two, such that e matches the first part of the sequence, and f matches the second part.[4]

For instance, if we want to capture executions which start with a session being opened, followed immediately by a deposit or withdrawal, we would write:[5]

```
[before UserInfo.openSession()] ;
  [after UserInfo.openSession()] ;
    ([before UserInfo.depositTo(..)] +
      [before UserInfo.withdrawFrom(..)]
    )
```

This regular expression matches if the first method call triggering events is that of `openSession`, and it is immediately followed either by a call to `depositTo` or `withdrawFrom`.

Repetition: Sometimes we want to capture the notion of a regular expression repeatedly matching a number of times in sequence. Given a regular expression e, we will write e* to match after any number of repetitions of e (including zero times). For example, `any*` would match with every event received, including right at the beginning before any events are seen.

For a practical use of the repetition operator, we can extend the previous example to capture a deposit or withdrawal occurring immediately after `openSession()` terminates:

```
any* ;
[after UserInfo.openSession()] ;
([before UserInfo.depositTo(..)] + [before UserInfo.withdrawFrom(..)])
```

[4] Note that the semicolon operator is being used as an infix operator and not as an expression terminator, as is the case in many programming languages.

[5] As in the case of the chapter on symbolic automata, we will initially write properties assuming a single user is interacting with the system at the time. We will relax this (unrealistic) constraint in due course.

Another example using repetition is to write a regular expression which captures deposits occurring outside the context of a session i.e. before any session is opened, or after a session is closed and before another is opened:

```
( [!after UserInfo.openSession()]* ;
    [before UserInfo.depositTo(..)]
) + (
  any*;
  [after UserInfo.closeSession()] ;
    [!before UserInfo.openSession()]* ;
      [before UserInfo.depositTo(..)]
)
```

The regular expression is split into two, matching if either (i) a deposit is received before a session is opened; or (ii) a deposit attempt follows an end of session event without another one being opened first.

Other basic expressions: We also include two expressions which, although not typically useful when writing a specification, will be used when we design and program our monitors.
The two additional monitoring elements are (i) the regular expression `nothing` which does not match any stream of events; and (ii) the regular expression `emptytrace` matches only the empty trace i.e. unless used as part of a bigger regular expression, it will only match at the beginning of the execution of the system before any event is observed.

$$
\begin{array}{rcl}
\langle \mathit{RegExp} \rangle & ::= & \texttt{nothing} \mid \texttt{emptytrace} \mid \texttt{any} \\
& \mid & \texttt{[}\langle \mathit{Event} \rangle\texttt{]} \mid \texttt{[!}\langle \mathit{Event} \rangle\texttt{]} \\
& \mid & \langle \mathit{RegExp} \rangle \ \texttt{+} \ \langle \mathit{RegExp} \rangle \\
& \mid & \langle \mathit{RegExp} \rangle \ \texttt{;} \ \langle \mathit{RegExp} \rangle \\
& \mid & \langle \mathit{RegExp} \rangle\texttt{*}
\end{array}
$$

Fig. 8.1 Monitoring regular expression syntax

The full syntax of monitoring regular expressions is given in Figure 8.1. Regular expressions are used to *match* sequences of events, but we have not discussed how we can use them to characterise bad behaviour. There are different ways of using regular expressions to write specifications — either using regular expressions to match bad traces or to match good ones. In most of this chapter we will be using the former, with specifications written as regular expressions which match if there is a violation. We will refer to these as *negative regular expressions*. We will discuss positive regular expressions later on in this chapter.

Recall Property 2:*"The transaction system must be initialised before any user logs in"*. This property is violated if a user logs in before the system is initialised, which can be written as a negative regular expression property in our tool as follows:

```
property "P2" bad behaviour {
  [!after BackEnd.initialise()]* ;
    [before UserInfo.openSession()]
}
```

This regular expression matches every time `openSession()` is called before the termination of `initialise()` is observed, allowing us to capture all violation instances of the property, each and every time one occurs. The string given after the `property` keyword is logged to the standard error stream whenever a violation is seen.

Exercise 8.1

Recall that in Exercise 7.1, we modified FiTS to add a new method `Backend.shutdown()`, such that the transaction system should no longer allow a session to open after a shutdown. Using a negative regular expression, express Property 2 but allowing `openSession()` only between initialisation and shutdown.

Express the following properties as monitoring regular expressions assuming that only a single user is accessing FiTS:

- Property 5: *"Once a user is disabled, he or she may not withdraw from an account until the administrator enables them again"*.
- Property 10: *"Logging can only be made to an active session (i.e. between a login and a logout)"*.

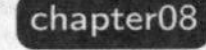

8.2 Monitoring Regular Expression Scripts

As in the case of EGCL and automata, code to parse and manipulate scripts with regular expressions is provided. The parser assumes a structure similar to EGCL and symbolic automata, with separate sections for `VERIFICATIONCODE`, `PRELUDE` and for the specifications tagged, in this case, `REGULAREXPRESSIONS`. A sample script checking Property 2 (*"The transaction system must be initialised before any user logs in"*) and the property that deposits can only take place during an open session, is shown below:

```
VERIFICATIONCODE

PRELUDE

package rv;
import fits.*;

REGULAREXPRESSIONS

property "P2" bad behaviour {
  [!after BackEnd.initialise()]* ;
    [before UserInfo.openSession()]
}

property "Deposits can only take place during an open session"
  bad behaviour {
  [!after UserInfo.openSession()]* ;
    [before UserInfo.depositTo(..)] +
  any*;
    [after UserInfo.closeSession()] ;
      [!before UserInfo.openSession()]* ;
        [before UserInfo.depositTo(..)]
}
```

Note that the order of operator precedence from highest to lowest is: repetition (*), sequential composition (;) and choice (+). For example, `[after A.a()] + [after B.b()];[after C.c()]*` would be interpreted as `[after A.a()] + ([after B.b()];([after C.c()]*))`.

Our tool will keep a representation of the regular expression specifications in the form of an abstract syntax tree, with the internal node corresponding to an operator (choice, sequential composition, repetition) and leaves corresponding to a basic regular expression (an event, a complemented event, `any`, `nothing`, `emptytrace`). A separate class for each type of internal and leaf node is provided, all of which inherit from an abstract class `RegExp` in the `structure` package:

- `RegExp.java` provides the abstract class which includes a number of methods, including (i) `String toString()` which returns the regular expression as a string; and (ii) `Set<Event> getRelevantEvents()` which returns the events referred to in the regular expression (including ones which are complemented).
- `Any.java`, `Nothing.java` and `EmptyTrace.java` all inherit from the `RegExp` abstract class, implementing the methods from that class.
- `MatchEvent.java` and `MatchEventComplement.java` are used to represent matching and complemented events (represented by `Event.java`) respectively, where an event is a method signature plus the optional before/after modality.

- `Choice.java` and `SequentialComposition.java` provide a class inheriting from `RegExp`, and in addition, methods to get the two constituent regular expression operands: `RegExp getLeft()` and `RegExp getRight()`.
- `Repetition.java` also inherits from `RegExp` and provides an additional method to get the regular expression appearing under repetition using `RegExp getChild()`.

Exercise 8.2

Familiarise yourself with the code provided, and do the following:

- Manually create objects representing the two regular expressions in the examples used in this chapter directly using the constructors. Check that they have been well constructed using the `toString()` method.
- Implement the `Set<Event> getRelevantEvents()` method in each child class. Confirm it works using the example properties.
- Implement a method `Boolean eventMatches(Event event)` in the `MatchEvent` and `MatchEventComplement` classes, which takes an event and checks whether or not it matches with the regular expression (based on whether it is a `before` or `after` event and on the class and method name, but ignoring parameter information). Recall that for simplicity we will assume that events in the regular expression (i) will all be class-specific i.e. we will not have events such as `*.initialise`; and (ii) will not contain overloaded methods with a different number of parameters or parameter types.

chapter08/MyRVTool

As in the previous chapters, regular expression parsing is provided in the accompanying code.

- `Main.java`: As in the case of EGCL and automata, this provides a simple class using the code in the files below to read a monitoring regular expression script and pretty prints it again to the standard output.
- `RegExpScript.java`: This class is used to represent, manipulate and handle monitoring regular expression scripts, and provides a constructor RegExpScript(String filename) which parses a monitoring regular expression script from a file (throwing an exception if a syntax error is found). The class provides a `String toString()` method to output the script back in text form.
  The class provides getter methods for the script: (i) the `VERIFICATIONCODE` section can be obtained using `String getVerificationCode()`; (ii) `String getPrelude()` returns the `PRELUDE` part; and (iii) `ArrayList<RegExpSpec>`

`getRegExpSpecs()` returns the regular expressions which are found in the `REGULAREXPRESSIONS` part of the script.

- `RegExpParser.java`: This contains parsing logic which need not concern us.

Exercise 8.3

Familiarise yourself with the additional code provided, and run it on the regular expression scripts for the different properties discussed in this chapter.

chapter08/MyRVTool

8.3 Verification of Monitoring Regular Expressions

The regular expressions we have seen in the previous section show how they allow us to express properties in a more structured manner. We will now move on to show how we can automatically obtain monitoring code from the regular expressions.

8.3.1 Regular Expressions and Automata

A well-known result in computer science is that regular expressions and finite state automata are equally expressive — for any finite state automaton, one can find an equivalent regular expression, and vice versa. Given that in this book we started with AspectJ specifications, then moved on to EGCL (which we translated to aspects), and then to automata (which we translated to EGCL), why not use this result to translate regular expressions into automata?

One can find various algorithms in the literature on how to translate regular expressions into automata. Consider the regular expression stating that observing either of the following initial behaviours is a violation (i) an administrator blacklisting a user immediately after a reconciliation at startup of accounts; (ii) a greylisting of a user immediately after a reconciliation at startup. This can be written as the following negative regular expression:

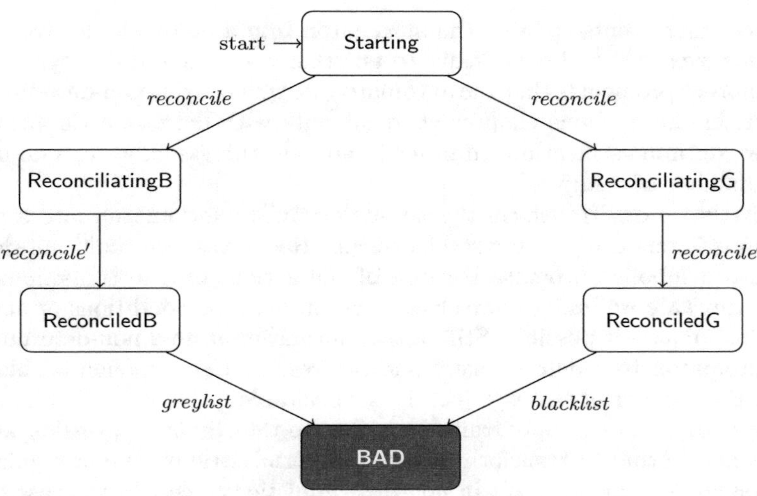

Fig. 8.2 An automaton obtained from the regular expression stating that neither blacklisting nor greylisting should be performed by an administrator immediately after a reconciliation of accounts. The tags on the transitions refer to the following events: (i) *blacklist* and *greylist* refer to `before TransactionSystem.ADMIN_blacklist(..)` and `before TransactionSystem.ADMIN_greylist(..)` respectively; (ii) *reconcile* and *reconcile′* refer to `before TransactionSystem.ADMIN_reconcile()` and `after TransactionSystem.ADMIN_reconcile()` respectively.

```
(
  [before TransactionSystem.ADMIN_reconcile()] ;
    [after TransactionSystem.ADMIN_reconcile()] ;
      [before TransactionSystem.ADMIN_blacklist(..)]
) + (
  [before TransactionSystem.ADMIN_reconcile()] ;
    [after TransactionSystem.ADMIN_reconcile()] ;
      [before TransactionSystem.ADMIN_greylist (..)]
)
```

A naïve translation of this property into a finite state automaton would produce the automaton shown in Figure 8.2. However, we can identify a number of problems with the process and the resulting automaton.

1. Firstly, recall that our automata had the peculiarity that they consume events not matching any outgoing transition, remaining in the same state. For instance, if a call to a method other than `ADMIN_reconciliate()` is observed in the initial state, the automaton will remain in the same state, and possibly capture a violation if a reconciliation and blacklisting follows. However, such a trace should not be tagged as a violation according to the regular expression we started off with. In order to avoid this issue we would need to add multiple other transitions and states to ensure that

any other events redirect the automaton into a non-violable state. This adds overhead and complexity to an otherwise simple property.[6]

2. Another problem is that the automaton we generated is non-deterministic, but in the previous chapter we dealt only with deterministic automata for runtime verification. In order to address this issue, we can do one of a number of things.
   Firstly we can transform the non-deterministic automaton into a deterministic one using standard algorithms. Recall that we could not do this with symbolic automata because of the actions on the transitions, but in this case we limited ourselves to events with no conditions or actions, which makes it possible. Still, a transformation from a non-deterministic automaton to a deterministic one can result in an exponential blow-up in the number of states, which is not desirable.
   Secondly, we can apply transformations to the regular expression we will be monitoring to transform it into a deterministic one i.e. a regular expression can never result in non-deterministic choices. In the case of the example we are using, this may be done by rewriting the property as follows:

```
[before TransactionSystem.ADMIN_reconcile()] ;
[after TransactionSystem.ADMIN_reconcile()] ;
([before TransactionSystem.ADMIN_blacklist(..)] +
  [before TransactionSystem.ADMIN_greylist (..)])
```

   Thirdly, we can decide to limit ourselves to deterministic regular expressions, just as we did with symbolic automata.
   The undesirable combination of non-determinism and symbolic actions is why we limited our regular expressions to use events but no conditions or actions. Whereas in automata it is easier for an engineer to check whether transitions are deterministic (they have to compare events and conditions on transitions exiting the same state) it is less straightforward with regular expressions (they would first have to deduce which events and conditions can fire concurrently before checking for mutual exclusivity), hence the choice.
3. The third problem is that in our symbolic automata we made our bad states sink states — once you reach a bad state you have no further transitions to follow and you remain indefinitely in that state. In contrast, with regular expressions we allowed for identifying violations multiple times. In order to address this we could either change symbolic automata to al-

[6] One way of handling this is to have a catch-all transition from every state, which matches any event not handled by the other transitions and leads to a sink state which never fails. However, this would require either adding many transitions (one for each other event which may occur) or making the automaton non-deterministic and allowing this catch-all transition to trigger only if none of the others do. Neither solution is ideal.

low for multiple violation flagging, or changing the meaning of monitoring regular expressions to capture only the first violation.

Addressing these different issues is possible, but would complicate and increase the complexity of the verification code considerably, particularly in order to deal with non-determinism. We will thus look at another approach, using *derivatives*.

Exercise 8.4

Change the property used in this section to capture reconciliation immediately followed by blacklisting and greylisting at any time i.e. not just when the system starts. Design a symbolic automaton which captures the same property (except that it will only flag the first violation), and assess the difficulties in performing such a transformation automatically.

chapter08

8.3.2 Regular Expression Derivatives

Another way of matching regular expressions is through the use of *derivatives* or *residuals*. The key to this approach is that with every event we observe, we can modify the regular expression we are trying to match to obtain a new one — called the derivative (or residual) of the original regular expression with respect to the new observation. The new expression is the one we have to match from now on.

For example, if we start trying to match regular expression `[a] ; ([b]+[c])` (where `a`, `b` and `c` are events) and we observe event `a`, then the derivative remaining to be matched is `[b]+[c]`. If we now observe event `b`, the derivative regular expression is `emptytrace` (since it matches the first choice but it does not match the second). We now know that the original regular expression matches since the remaining regular expression accepts the empty trace of events. On the other hand, if we initially observed an event other than `a`, the derivative regular expression would be `nothing`, since the trace will never match.

Using this approach, with every incoming event we shave off parts of the regular expression until the remaining expression accepts the empty trace of events, in which case we report a match.

In order to implement this approach, we need two algorithms (i) an algorithm to check whether a regular expression matches the empty trace; and (ii) an algorithm which, given an event and a regular expression, returns the resulting derivative — the regular expression which corresponds to what is still to be matched after the event just received. Once we have these two components, we can easily monitor the regular expression. We create a monitoring variable initially set to the regular expression to be matched, and every time an event is observed we update this variable to be the derivative using algorithm (ii). At each stage we also check whether the empty trace can be accepted by the regular expression in the variable using algorithm (i), and if it does, we flag violation.

Let us start with the first algorithm — whether a given regular expression matches the empty trace. We would like to define a function *hasEmpty*(e) which returns true if and only if e matches the empty trace. The definition is straightforward for most of the cases: (i) regular expression `emptytrace` matches the empty trace; (ii) regular expression `nothing` does not match the empty trace; (iii) regular expressions `[a]`, `[!a]` and `any` do not contain the empty trace; and (iv) regular expression e`*` matches the empty trace for any regular expression e (repeating e zero times). That leaves the choice and sequential composition operators: e_1 `+` e_2 matches the empty trace if at least one of e_1 and e_2 match the empty trace, while e_1 `;` e_2 matches the empty trace if both e_1 and e_2 match the empty trace. The resulting algorithm is given below[7]:

$$
\begin{aligned}
hasEmpty(\texttt{nothing}) &\stackrel{df}{=} \mathit{false}\\
hasEmpty(\texttt{emptytrace}) &\stackrel{df}{=} \mathit{true}\\
hasEmpty(\texttt{any}) &\stackrel{df}{=} \mathit{false}\\
hasEmpty(\texttt{[a]}) &\stackrel{df}{=} \mathit{false}\\
hasEmpty(\texttt{[!a]}) &\stackrel{df}{=} \mathit{false}\\
hasEmpty(e_1 \texttt{ + } e_2) &\stackrel{df}{=} hasEmpty(e_1) \text{ or } hasEmpty(e_2)\\
hasEmpty(e_1 \texttt{ ; } e_2) &\stackrel{df}{=} hasEmpty(e_1) \text{ and } hasEmpty(e_2)\\
hasEmpty(e\texttt{*}) &\stackrel{df}{=} \mathit{true}
\end{aligned}
$$

Exercise 8.5

Declare a method `Boolean hasEmpty()` in the `RegExp` abstract class and, following the definition of the function *hasEmpty*(), and implement the method in the classes inheriting from this class.

chapter08/MyRVTool

As for the second algorithm which, given an event `a` and a regular expression e, returns the resulting regular expression derivative, we define the function *derivative*(`a`, e). If *derivative*($\texttt{a}_1$, e) returns regular expression e', a trace

[7] We use the symbol $\stackrel{df}{=}$ to denote a definition — with the term on the left being defined to be the expression on the right.

of events ⟨a$_1$, a$_2$, a$_3$... a$_n$⟩ matches e if and only if trace ⟨a$_2$, a$_3$,... a$_n$⟩ matches e'. The definition is shown below:

derivative(a, nothing) $\stackrel{df}{=}$ nothing
derivative(a, emptytrace) $\stackrel{df}{=}$ nothing
derivative(a, any) $\stackrel{df}{=}$ emptytrace
derivative(a, [b]) $\stackrel{df}{=}$
$\begin{cases} \texttt{emptytrace} & \text{if } \texttt{a} = \texttt{b} \\ \texttt{nothing} & \text{otherwise} \end{cases}$
derivative(a, [!b]) $\stackrel{df}{=}$
$\begin{cases} \texttt{emptytrace} & \text{if } \texttt{a} \neq \texttt{b} \\ \texttt{nothing} & \text{otherwise} \end{cases}$
derivative(a, e_1 + e_2) $\stackrel{df}{=}$ *derivative*(a, e_1) + *derivative*(a, e_2)
derivative(a, e_1 ; e_2) $\stackrel{df}{=}$
$\begin{cases} \textit{derivative}(\texttt{a}, e_1) \ ; \ e_2 & \text{if not } \textit{hasEmpty}(e_1) \\ \textit{derivative}(\texttt{a}, e_1) \ ; \ e_2 \ + \ \textit{derivative}(\texttt{a}, e_2) & \text{otherwise} \end{cases}$
derivative(a, e*) $\stackrel{df}{=}$ *derivative*(a, e) ; e*

The first two lines of the definition are quite straightforward — after consuming an event, neither nothing nor emptytrace can match again, resulting in the derivative nothing. The next three lines take into consideration matching with single event regular expressions. If it matches, then the derivative is emptytrace, and if it does not, the derivative is nothing. The derivative of a choice can be calculated by simply taking the derivative of both operands of the choice operator.

The derivative of the sequential composition of two regular expressions is more complex and is split into two parts: (i) if the first operand does not match the empty trace, we take the derivative of the first regular expression in the sequential composition leaving the second unchanged; but (ii) if the first operand matches the empty trace we must also must start consuming the second operand.

Finally, the derivative of the repetition operator can be derived by expanding e* to emptytrace + (e ; e*), allowing us to compute the derivative according to the rules for choice and sequential composition, simplifying the result using standard regular expression equivalence rules (e.g. nothing + $e \equiv e$).

Exercise 8.6

In the RegExp abstract class, add a method RegExp derivative(Event e). Define this method in the classes representing the regular expression elements and operators according to the *derivative* function defined above.

chapter08/MyRVTool

You will have noticed that to solve the exercise above, one must match an event with another. As long as events are concrete ones i.e. they do not use * to match generic classes, this is simple to do, but once we allow for generic events and events that can be overloaded (i.e. whether they match or not may depend on the number and type of parameters), the implementation becomes more intricate. This explains why we have limited ourselves to concrete events without overloading in this chapter.

Now that we have these two algorithms, we can define how we can build a runtime monitor and verifier from a regular expression. For a single regular expression e, the overall algorithm uses an EGCL script as follows:

- Create a new static `RegExp` monitoring state variable `currentRegExp` and initialise it to e, the regular expression to be matched. We will also add a new Boolean variable `triggered` to know when a rule has already triggered (initially set to *false*).
- For each event which e refers to i.e. the events that are returned by e`.getRelevantEvents()`, create a rule which updates `currentRegExp` with the result of calling the `derivative` method with this event and the current value of `currentRegExp`. For instance, in order to handle event `before UserInfo.depositTo(..)` one would generate the following (partial) rule:[8]

```
before UserInfo.depositTo(..) |
  -> {
    Event event = new Event("before", "UserInfo.depositTo");
    currentRegExp = currentRegExp.derivative(event);
    triggered = true;
  }
```

- Any event not referred to by the regular expression should also trigger a rule. In order to do so, we will generate a general rule similar to the one above but using the derivative on an event with a spurious name ("#"). Since we do not want events we handled earlier to trigger again with this catch-all rule, we guard it by a variable to check whether a rule has already been triggered:

[8] Note that we use the `Event` constructor which takes the modality (before or after) and the name of a method, both as strings, and produces a matching event with generic parameter matching.

```
before *.*(..) | !triggered
  -> {
    Event event = new Event("before", "#");
    currentRegExp = currentRegExp.derivative(event);
    triggered = true;
  }

after *.*(..) | !triggered
  -> {
    Event event = new Event("after", "#");
    currentRegExp = currentRegExp.derivative(event);
    triggered = true;
  }
```

- Finally, at the end, we reset the `triggered` variable in preparation for the next event:

```
before *.*(..) | -> { triggered = false; }

after *.*(..) | -> { triggered = false; }
```

Every time `currentRegExp` is updated, we check whether it can accept the empty string, in which case it will trigger a violation report. We can do so by replacing the catch-all rules above to check whether a violation occurred:

```
before *.*(..) |
  -> {
    if (currentExp.containsEmpty()) {
      Assertion.alert("Violation detected");
    }
    triggered = false;
  }

after *.*(..) |
  -> {
    if (currentExp.containsEmpty()) {
      Assertion.alert("Violation detected");
    }
    triggered = false;
  }
```

This check would also have to be performed just after `currentRegExp` is initialised.

To handle multiple regular expression properties specified in the same script we would simply have to (i) create separate monitoring state variables, one for each property; and (ii) create the EGCL rules for all regular expressions to be monitored.

Exercise 8.7

1. Create a method `String toEGCL()` in the `RegExpScript` and `RegExp` class, which will generate the EGCL script for a single regular expression property as explained above. Change `Main.java` to output the EGCL script as a text file.
2. Extend the code to report the error string given in the regular expression script after the `property` keyword (and stored in the `RegExpScript` object) instead of a generic violation error report.
3. Modify `Main.java` to use the `EGCLScript.toAspectJ()` method and to generate an AspectJ file.
4. Extend the solutions above to work for multiple properties given in a regular expression script. Modify `Main.java` accordingly.

chapter08-MyRVTool

8.4 Regular Expressions as Positive Specifications

Until this point, we have used regular expressions to specify undesirable behaviour, hence the `bad behaviour` keywords in our scripts. However, other approaches are possible and in some cases perhaps even more effective in writing specifications. In this section, we will consider using regular expressions as *positive* specifications, stating how we expect the system to behave, rather than how we expect it *not* to behave.

Recall, Property 2: *"The transaction system must be initialised before any user logs in"*, for which we wrote the following negative specification:

```
property "P2" bad behaviour {
  [!after BackEnd.initialise()]* ;
    [before UserInfo.openSession()]
}
```

An alternative way to express this property is to describe good behaviour, and write a monitoring regular expression which matches only traces in which `initialise()` happens before `openSession()`. One way of expressing this property in a positive manner is the following script excerpt using the `good behaviour` rather than `bad behaviour` option:

```
property "P2" good behaviour {
  [!before UserInfo.openSession()]* ;
    [after BackEnd.initialise()] ;
      any*
}
```

As long as the behaviour observed can be extended to match the regular expression, we have not yet encountered a violation. Notice the difference from the negative regular expression approach in which a violation would match the regular expression instead.

As another example, consider the property that states that deposits must occur within the context of a session i.e. after a session is opened, but before it is closed. A positive version of this property is the following:

```
property "Deposit only during open session" good behaviour {
  ( [!before UserInfo.depositTo(..)]* ;
      [after UserInfo.openSession()] ;
        [!after UserInfo.closeSession()]* ;
          [after UserInfo.closeSession()]
  )*
}
```

Once again, as long as the observed behaviour can be extended to match the regular expression, there has not yet been a violation. For instance, assume that we have observed two withdrawals followed by the opening of a session as shown below:

```
before UserInfo.withdrawFrom(..)
after  UserInfo.withdrawFrom(..)
before UserInfo.withdrawFrom(..)
after  UserInfo.withdrawFrom(..)
before UserInfo.openSession()
after  UserInfo.openSession()
```

This trace has not yet deviated from the behaviour described in the regular expression since it can be extended to match the property, for example by adding the following events:

```
before UserInfo.closeSession()
after  UserInfo.closeSession()
```

So, are positive monitoring regular expressions more effective when writing a specification? In reality, it depends on the property being written. Some specifications are more readable as positive regular expressions, while others are better written as negative ones. What about efficiency of deciding whether there has been a violation? In the case of positive regular expressions we spoke about being able to extend the observed trace of events to match the regular expression which sounds computationally expensive. But is it?

Recall that while monitoring we keep track of the regular expression that still needs to be satisfied using the *derivative* function. It thus suffices to check whether there exist traces that would match with the remaining derivative. We can write a simple algorithm similar to the one we wrote to check whether a regular expression accepts the empty string or not, but to return whether a given regular expression is unsatisfiable:

$$
\begin{aligned}
\mathit{cannotMatch}(\texttt{nothing}) &\stackrel{df}{=} \mathit{true}\\
\mathit{cannotMatch}(\texttt{emptytrace}) &\stackrel{df}{=} \mathit{false}\\
\mathit{cannotMatch}(\texttt{any}) &\stackrel{df}{=} \mathit{false}\\
\mathit{cannotMatch}(\texttt{[a]}) &\stackrel{df}{=} \mathit{false}\\
\mathit{cannotMatch}(\texttt{[!a]}) &\stackrel{df}{=} \mathit{false}\\
\mathit{cannotMatch}(e_1\ \texttt{+}\ e_2) &\stackrel{df}{=} \mathit{cannotMatch}(e_1) \text{ and } \mathit{cannotMatch}(e_2)\\
\mathit{cannotMatch}(e_1\ \texttt{;}\ e_2) &\stackrel{df}{=} \mathit{cannotMatch}(e_1) \text{ or } \mathit{cannotMatch}(e_2)\\
\mathit{cannotMatch}(e\star) &\stackrel{df}{=} \mathit{false}
\end{aligned}
$$

The definitions above should be rather straightforward to justify and understand. Clearly, `nothing` will never be satisfied (no matter what events we are still to see), but `emptytrace`, `any` and the event matching `[a]` and `[!a]` can all be satisfied (the first by the empty trace, the others by an appropriate single event). As to the constructors: repetition is always satisfiable (by the empty trace), choice cannot be satisfied if both its operands cannot be satisfied, and sequential composition cannot be satisfied if at least one of its operands cannot be satisfied.

Using this algorithm, we can transform positive regular expressions using EGCL rules just as before but changing the violation condition to match when the *cannotMatch* function returns true. The catch-all EGCL rules added to reset the `triggered` variable and to check for violations would be changed as follows:

```
before *.*(..) |
  -> {
    if (currentExp.cannotMatch()) {
      Assertion.alert("Violation detected");
    }
    triggered = false;
  }

after *.*(..) |
  -> {
    if (currentExp.cannotMatch()) {
      Assertion.alert("Violation detected");
    }
    triggered = false;
  }
```

In practice our tool will support both positive and negative regular expressions, thus allowing the person writing the specification to choose the most appropriate notation.

Exercise 8.8

1. Declare a `Boolean cannotMatch()` method as part of the `RegExp` class and implement it in the classes that inherit from it.
2. Following the approach used so far for negative, i.e. bad behaviour, regular expressions, change the violation detection condition as explained in this section to handle the positive counterparts.

chapter08/MyRVTool

8.5 Parameterised Regular Expression Monitoring

Just as we did in the previous chapters, we will now look at lifting regular expressions to work on separate instances of a class independently. We have already expressed a property using a regular expression which should be parameterised — if we want to ensure that deposits can only take place while the user has an open session, we would have to differentiate between calls made by different users. If a user has an open session, other users should still not be allowed to perform transfers. However, the way we expressed this property does not capture this nuance, and we assumed that only a single user would have to be using the system (or be monitored) for the property to work as written:

```
property "Deposits can only take place during an open session"
  bad behaviour {
  [!after UserInfo.openSession()]* ;
    [before UserInfo.depositTo(..)] +
  any*;
    [after UserInfo.closeSession()] ;
      [!before UserInfo.openSession()]* ;
        [before UserInfo.depositTo(..)]
}
```

We will follow the approach used in previous chapters to allow for quantification of such properties to function on a per-object basis using a `foreach` construct similar to the ones used for EGCL and automata:

```
property "Deposits can only take place during an open session"
  foreach target UserInfo
  bad behaviour {
  [!after UserInfo.openSession()]* ;
    [before UserInfo.depositTo(..)] +
  any*;
    [after UserInfo.closeSession()] ;
      [!before UserInfo.openSession()]* ;
        [before UserInfo.depositTo(..)]
}
```

Note that we do not support adding new monitoring state (as we did before using the `keep` keyword) since regular expressions do not support conditions and actions. For the same reason, we do not name the instance of the target as we did with EGCL and automata. However, just as before, the specification will instrument a monitor for each `UserInfo` object and match them individually against the property. As before, the methods we monitor in the parameterised property must all target the same class which we quantify over.

Exercise 8.9

Write parameterised regular expression specifications for the following properties:

- Property 5: *"Once a user is disabled, he or she may not withdraw from an account until the administrator enables them again"*.
- Property 10: *"Logging can only be made to an active session (i.e. between a login and a logout)"*.

chapter08-parameterised

In order to replicate the monitor for the regular expression, we can use the same approach that we used for parameterised automata and use the parameterisation in EGCL. The rules we showed in Section 8.3.2 would be included under a `foreach` construct in the generated EGCL. For instance, the property requiring deposits to be made during an open session would be transformed into the following:

```
property foreach target (UserInfo _t)
  keep (
    RegExp currentRegExp defaultsTo new RegExp(...),
    Boolean triggered defaultsTo false
  ){

  // Trigger for all relevant events
  after UserInfo.openSession() |
    -> {
      Event event = new Event("before", "UserInfo.depositTo");
      currentRegExp = currentRegExp.derivative(event);
      triggered = true;
    }

  ...

  // Trigger with any other event
  before UserInfo.*(..) | !triggered
    -> {
      Event event = new Event("before", "#");
      currentRegExp = currentRegExp.derivative(event);
      triggered = false;
    }

  ...
}
```

Apart from adding the `foreach` construct, we add the monitoring variable local to the replicated property (whose name would be automatically generated), effectively generating a variable for each `UserInfo` object. This is initialised to the regular expression we want to monitor (written `new RegExp(...)` in the script above, but would have to be set to the regular expression in question). The only other change we have to do is to modify the final catch-all rule, limiting it to methods targeting `UserInfo` objects i.e. `UserInfo.*(..)` rather than `*.*(..)` as we did before.

To compile parameterised regular expressions, we will use the same strategy as we did before — we produce parameterised EGCL script object, and transform it into AspectJ. The support code for handling parameterised regular expressions can be found in chapter08-parameterised under the `MyRVTool` folder. In addition to the code we saw earlier for parameterised regular expressions we have the following new classes:

- `RegExpScript.java` provides a method returning the parameterised regular expressions found in the script: `ArrayList<ParameterisedRegExpSpec> getParameterisedRegExpSpecs()`.
- `ParameterisedRegExpSpec.java` is used to represent parameterised regular expressions with the following getters: (i) `String getTargetType()` returns the type of the target; and (ii) `RegExpSpec getRegExpSpec()` returns the regular expression.

Exercise 8.10

Review the code provided and then:

1. Modify the code you wrote earlier to monitor negative regular expressions so as to handle negative parameterised regular expressions.
2. Do the same to compile positive parameterised regular expressions.
3. Change `Main.java` to read a regular expression script file (which may use parameterisation) and output an AspectJ file and the verification class Java file.
4. Test the tool using the properties from the previous exercise.

chapter08-parameterised/MyRVTool

8.6 Conclusions

In this chapter, we have presented regular expressions as a more structured notation than automata. Their inherent structure makes them compositional and thus more easily reusable. This comes at a cost since the monitoring and verification algorithms are now substantially more complex.

It is worth noting that the choice of property specification notation is an important one that depends largely on the properties that one needs for the system at hand. For some properties, particularly simple ones, symbolic automata provide an excellent and visual notation. For other properties, regular expressions or other compositional notations may be more appropriate. In the next chapter, we will look at another specification logic, Linear Temporal Logic.

Chapter 9
Linear Temporal Logic

In the previous chapter we showed how regular expressions can be used to express temporal properties of systems in a compositional manner. In computer science, various *temporal logics* have been developed to allow for such specifications. Linear Temporal Logic[1] (LTL) is one such logic which was originally proposed in 1977 by Amir Pnueli as a logic to specify temporal properties of programs, but has since been shown to be particularly well suited as a logic for automated system verification, particularly using static analysis, but also for runtime verification. In this chapter we will be introducing LTL, showing how it can be used to express properties, and how LTL properties can be monitored.

9.1 Syntax and Semantics

LTL formulae express properties about infinite traces of events, expressing how the system should evolve over a particular run. This may not sound all that well suited for runtime verification which deals only with *finite* observable traces. However, as we shall see, given a finite observed trace one can still check whether or not a property has already been violated.[2]

As in the case of regular expressions, the basic building block is the notion of an event such as: `[before UserInfo.openSession()]`, which is satisfied if the first observed event matches i.e. the first thing observed is a call to the method to open a new session.

[1] Different temporal logics allow for different aspects of temporal behaviour one would like to express properties of. One such aspect is whether time is to be seen as a linear chain of events or as a branching tree of possible futures. The former is useful when we would like to specify properties about behaviour in a particular run or all runs of a system, the latter when we want to express properties about possible behaviour. As the name suggests, Linear Temporal Logic uses a linear model of time and is of the former kind.

[2] This is not the case of all LTL formulae, but we will come back to this later on.

C. Colombo, G. J. Pace, *Runtime Verification*,
https://doi.org/10.1007/978-3-031-09268-8_9

$$
\begin{array}{rcl}
\langle LTL \rangle & ::= & \texttt{true} \mid \texttt{false} \mid \texttt{[}\langle Event \rangle\texttt{]} \\
& \mid & \texttt{not}\ \langle LTL \rangle \\
& \mid & \langle LTL \rangle\ \texttt{and}\ \langle LTL \rangle \\
& \mid & \langle LTL \rangle\ \texttt{or}\ \langle LTL \rangle \\
& \mid & \langle LTL \rangle\ \texttt{implies}\ \langle LTL \rangle \\
& \mid & \texttt{next}\ \langle LTL \rangle \\
& \mid & \texttt{globally}\ \langle LTL \rangle \\
& \mid & \texttt{finally}\ \langle LTL \rangle \\
& \mid & \langle LTL \rangle\ \texttt{until}\ \langle LTL \rangle
\end{array}
$$

Fig. 9.1 LTL syntax

LTL formulae can also be combined using conjunction (written e_1 `and` e_2), disjunction (written e_1 `or` e_2), implication (written e_1 `implies` e_2) and negation (written `not` e). For example, the property which says that the first event observed must be to either deposit or withdraw from a user account can be written in LTL as follows:

```
[before UserInfo.withdrawFrom(..)] or [before UserInfo.depositTo(..)]
```

As the name of the logic indicates, LTL provides a number of operators to express temporal properties — constraints regarding the expected order of the events. A trace satisfies the formula `next` e (where e is any LTL formula) if and only if, after dropping the first event of the trace, the remaining trace of events satisfies formula e. So, if we want to write the property which says that if the first event is a request to transfer funds at the user interface level, then the next (second) event should be a call to withdraw funds (from the source account) at the user object level even before the request to perform the transfer terminates, one would write:

```
[before FrontEnd.USER_transferToOtherAccount(..)] implies
  next [before UserInfo.withdraw(..)]
```

The formula `finally` e says that if we wait long enough (ignore sufficiently many events), then the LTL formula e will eventually hold. For example, we can change the property above to specify that if the first event is a request to transfer funds at the interface level, then eventually we should see a call to withdraw funds at the account level, after which we should eventually see a call to deposit funds also at the account level:

```
[before FrontEnd.USER_transferToOtherAccount(..)] implies
  finally (
    [before UserAccount.withdrawFrom(..)] and
    finally [before UserAccount.depositTo(..)]
  )
```

Note that by nesting the `finally` operators we ensure that the order of the withdrawal and deposit is as specified. Had we separated the operators, they would have been allowed in either order:[3]

```
[before FrontEnd.USER_transferToOtherAccount(..)] implies
  ((finally [before UserAccount.withdrawFrom(..)]) and
   (finally [before UserAccount.depositTo(..)])
  )
```

The formula `globally` e indicates that LTL formula e should hold at all points in time starting now and later. For example, we can express the property that the administrator never reconciles accounts twice in a row using an LTL formula as follows:

```
globally (
  [after FrontEnd.ADMIN_reconcile()] implies
    not (next [before FrontEnd.ADMIN_reconcile()])
)
```

We can also extend the property imposing an order on the withdrawal and the deposit performed during a transfer to ensure that the order is respected whenever a transfer is requested and not just if it is called immediately upon start up of the system:

```
globally (
  [before FrontEnd.USER_transferToOtherAccount(..)] implies
    finally (
      [before UserAccount.withdrawFrom(..)] and
      finally [before UserAccount.depositTo(..)]
    )
  )
)
```

[3] In reality, the order would be desirable, since in financial transaction systems, an important rule of thumb is to always withdraw before you deposit, ensuring that any pending balance is always in the system's favour.

Finally, the formula e_1 `until` e_2 holds if (i) LTL formula e_1 holds until e_2 becomes true, and (ii) e_2 will eventually become true[4]. For instance, to specify that users may not open a session until the system is initialised, and that the system must eventually be initialised, can be expressed in LTL as follows:

```
not [before UserInfo.openSession()] until
  [before BackEnd.initialise()]
```

For instance, we can specify that during the execution of a withdrawal, a call to deposit may not be made:

```
globally (
  [before UserInfo.withdrawFrom(..)] implies
    (not [before UserInfo.depositTo(..)] until
      [after UserInfo.withdrawFrom(..)]
    )
)
```

In addition, LTL also includes the basic formulae `true` and `false` corresponding to the property which is always satisfied and the one which is always violated respectively.

The full syntax of LTL is given in Figure 9.1. The order of operator precedence of the operators, from highest to lowest is: unary operators (i.e. `not`, `next`, `finally` and `globally`), `until`, `implies`, `and` and finally `or`.

It is worth noting that in all the examples above we used positive properties (specifying what we expect the system to do), rather than negative ones (specifying what wrong executions would look like). This is the way LTL is typically used to write specifications, and in the rest of the chapter we will limit ourselves to positive specifications i.e. equivalent to `good behaviour` specifications used in regular expressions.

Exercise 9.1

Explain the meaning of the following four LTL formulae:

```
not [before BackEnd.shutdown()] until
  [after BackEnd.initialise()]

globally (
  [after UserInfo.makeGreylisted()] implies
```

[4] In temporal logic sometimes we use two different *until* operators, a strong one which requires that e_2 will eventually hold (corresponding to the one we are using) and a weak one in which this is not required. We will only be using strong until in this chapter.

```
  next [before UserInfo.removeGoldStatus()]
)

globally (
  [after UserInfo.makeBlacklisted()] implies (
    not [before UserInfo.transfer(..)] until
      [after UserInfo.makeWhitelisted()]
  )
)

globally (
  [after UserInfo.makeBlacklisted()] implies (
    not [before UserInfo.transfer(..)] until (
      [after UserInfo.makeWhitelisted()] and (
        not [before UserInfo.transfer(..)] until
          [after UserInfo.openSession()]
      )
    )
  )
)
```

Assuming that the system handles only one user at a time, write the following properties in LTL:

- There should never be two whitelist calls in sequence.
- After being blacklisted, a user must be disabled before they are whitelisted again.
- Once blacklisted, a user may not perform any transfers until they are whitelisted.
- A request to open a session must be terminated either by a request to close the session or a blacklist event, before another request to open a session is received.
- A withdrawal and deposit must be performed during a call to `USER_transferToOtherAccount`.

9.2 Parsing LTL Formulae

As we have done in previous chapters, we provide code to parse, represent and analyse LTL properties. The syntax used for LTL properties follows that used in the previous section[5]. Properties are specified in a similar way as we did with regular expressions, prefixed with the `property` keyword and a property name. Although we only allow positive properties in LTL, we will keep the `good behaviour` keywords to make the meaning more explicit.

Recall Property 2: *"The transaction system must be initialised before any user logs in"*. We can express this in an LTL property as follows:

```
property "P2" good behaviour {
  not [before UserInfo.openSession] until
    [after BackEnd.ADMIN_initialise]
}
```

Exercise 9.2

Express the following properties as LTL formulae assuming that only a single user is accessing FiTS:

- Property 5: *"Once a user is disabled, he or she may not withdraw from an account until the administrator enables them again"*.
- Property 10: *"Logging can only be made to an active session (i.e. between a login and a logout)"*.

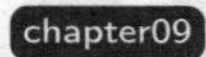

LTL scripts are organised in the same way as EGCL, automata and regular expression scripts. They are split into three sections: a `VERIFICATIONCODE`, a `PRELUDE` and an `LTL` section. A full LTL script for checking Property 2 (*"The transaction system must be initialised before any user logs in"*) is shown below:

```
VERIFICATIONCODE

PRELUDE
package rv;
import fits.*;

LTL
property "P2" good behaviour {
  not [before UserInfo.openSession] until
    [after BackEnd.initialise()]
}
```

[5] In most of the literature using LTL, one finds symbolic operators syntactically different from the ones we use, but we adopted this notation to make the transition to writing specifications in a machine readable format easier.

Just as in the case of regular expressions, LTL formulae are represented as abstract syntax trees, with different classes for the basic LTL formulae and constructors, all of which inherit from the abstract class `LTL`. The content of the classes follows closely that of regular expressions:

- `LTL.java` provides the abstract class from which all the LTL abstract syntax tree node classes inherit. The class provides the following methods: (i) `String toString()` returns the LTL formula as a string; and (ii) `Set<Event> getRelevantEvents()` which returns a set of the events appearing in the LTL formula.
- `True.java` and `False.java` inherit from the `LTL` class, providing an implementation of the methods for these LTL formula instances.
- `MatchEvent.java` represents an event and provides the method `Event getEvent()` which returns the matching event.
- `Not.java`, `Next.java`, `Finally.java` and `Globally.java` provide the concrete classes for the unary operators, all of which include the method `LTL getChild()`.
- `And.java`, `Or.java`, `Implies.java` and `Until.java` each provide a class inheriting from `LTL`, and provides methods to get the two operands: `LTL getLeft()` and `LTL getRight()`.

Exercise 9.3

Familiarise yourself with the code provided, then program the following:

- Create objects representing LTL formulae of your choice from examples and exercises in this chapter. Check that they are well-defined using the `toString()` method.
- Implement the `Set<Event> getRelevantEvents()` method in each child class and check that it works using the example properties.
- Implement a method `Boolean eventMatches(Event event)` in the `MatchEvent` class. This method will take an observed event as a parameter and returns whether the two match. As in the case of regular expressions, for simplicity we will assume that events (i) are all class-specific i.e. we will not have events such as `*.initialise`; and (ii) will not contain overloaded methods with a different number of parameters or parameter types.

chapter09/MyRVTool

The code provided for LTL scripts, covering parsing and manipulation of LTL formulae is organised as follows:

- `Main.java`: As in previous chapters, a class is provided to read an LTL script and pretty print it again to the standard output.

- `LTLScript.java`: This class is used to handle LTL scripts, and provides a constructor `LTLScript(String filename)` which parses such a script from a file. The class also provides a `String toString()` method to output the script back in text form.
  Getter methods for the parts of the script are provided as follows: (i) the `VERIFICATIONCODE` section can be accessed using the method `String getVerificationCode()`; (ii) the `PRELUDE` using `String getPrelude()`; and (iii) the LTL formulae as given in the `LTL` part of the script using `ArrayList<LTL> getLTLSpecs()`.
- `LTLParser.java`: This contains additional parsing logic which we will not need to delve into.

Exercise 9.4

Familiarise yourself with the code provided and modify `Main.java` to print the relevant events for each LTL formula given in a script.

chapter09/MyRVTool

9.3 LTL Runtime Verification

As was already mentioned, LTL properties are about infinite traces. It suffices to look at the LTL property `finally [a]`. Just because we have observed a thousand events but none of them being the event `a` does not mean that the property is violated. No matter how long the observation trace is, as long as it is finite we cannot have evidence of a violation.[6] On the other hand, violation of certain LTL formulae can be decided with a finite trace. For example, given the property `globally [a]`, we have evidence of a violation as soon as we observe an event which does not match the event `a`. No matter what happens in the future, the property is guaranteed to be violated.

Our aim is to build a monitoring algorithm which will flag a violation as soon as one has enough evidence to show that no matter what will happen in the future, the property has been violated once and for all. For example, consider the property which says that greylisting a user should be followed immediately by making them a *normal user* i.e. the user loses gold or silver status. In LTL this can be written as follows:

[6] There are also LTL properties for which neither evidence of satisfaction nor violation can be given through a finite trace. Consider the property `globally (finally [a])`, which says that event `a` must happen infinitely often. No matter what finite trace we observe, we cannot conclude anything about the system with respect to this property.

```
globally (
  [after UserInfo.greylist()] implies
    next [before UserInfo.makeNormalUser()]
)
```

If the first four events we observe are the following, we have no evidence of a property violation:

```
before UserInfo.openSession()
after UserInfo.openSession()
before UserInfo.greylist()
after UserInfo.greylist()
```

Similarly, no violation is detected if next event we see were to be `before UserInfo.makeNormalUser()`. However, if the next event instead was `before UserInfo.openSession()`, no matter what the infinite future holds for us, the property has been violated. This is the approach that we will be encapsulating in the verification algorithm.

One can find different verification algorithms for LTL in the literature, but in this chapter we will focus on a derivative-based approach. Just as we did with regular expressions, we will start with a means of computing the derivative of an LTL formula with respect to an event:

$$
\begin{aligned}
&\textit{derivative}(\texttt{a, true}) && \stackrel{df}{=} \texttt{true} \\
&\textit{derivative}(\texttt{a, false}) && \stackrel{df}{=} \texttt{false} \\
&\textit{derivative}(\texttt{a, [b]}) && \stackrel{df}{=} \begin{cases} \texttt{true} & \text{if } \texttt{a}=\texttt{b} \\ \texttt{false} & \text{otherwise} \end{cases} \\
&\textit{derivative}(\texttt{a, not } e) && \stackrel{df}{=} \texttt{not}(\textit{derivative}(\texttt{a}, e)) \\
&\textit{derivative}(\texttt{a}, e_1 \texttt{ and } e_2) && \stackrel{df}{=} \\
&\qquad \textit{derivative}(\texttt{a}, e_1) \texttt{ and } \textit{derivative}(\texttt{a}, e_2) \\
&\textit{derivative}(\texttt{a}, e_1 \texttt{ or } e_2) && \stackrel{df}{=} \\
&\qquad \textit{derivative}(\texttt{a}, e_1) \texttt{ or } \textit{derivative}(\texttt{a}, e_2) \\
&\textit{derivative}(\texttt{a}, e_1 \texttt{ implies } e_2) && \stackrel{df}{=} \\
&\qquad \textit{derivative}(\texttt{a}, e_1) \texttt{ implies } \textit{derivative}(\texttt{a}, e_2) \\
&\textit{derivative}(\texttt{a, next } e) && \stackrel{df}{=} e \\
&\textit{derivative}(\texttt{a, finally } e) && \stackrel{df}{=} \textit{derivative}(\texttt{a}, e) \texttt{ or finally } e \\
&\textit{derivative}(\texttt{a, globally } e) && \stackrel{df}{=} \textit{derivative}(\texttt{a}, e) \texttt{ and globally } e \\
&\textit{derivative}(\texttt{a}, e_1 \texttt{ until } e_2) && \stackrel{df}{=} \\
&\qquad \textit{derivative}(\texttt{a}, e_2) \texttt{ or } \textit{derivative}(\texttt{a}, e_1 \texttt{ and } (e_1 \texttt{ until } e_2))
\end{aligned}
$$

The first two lines of the definition are straightforward. Once we hit `true` or `false`, no matter what we observe, the remaining formula to match will not change. The next line handles a single event LTL formula — if it matches, then the formula is satisfied and the derivative is `true`, otherwise we have just witnessed a violation and the derivative is `false`.

The next four lines handle the derivative of the Boolean operators (negation, conjunction, disjunction and implication), which simply distributes over the operators.

The derivative of a `next` formula simply drops the operator. The operator `globally` e holds if and only if e `and globally` e. Applying the definition of the derivative of conjunction and the `next` operator, we obtain the derivative formula shown. Similarly, `finally` e is equivalent to e `or finally` e, and applying the rules we have seen, we obtain the derivative as shown above. The definition of the derivative of the `until` operator can be similarly derived.

Exercise 9.5

Implement the `LTL derivative(Event event)` method in the classes inheriting from the `LTL` abstract class.

chapter09/MyRVTool

As we have seen, LTL is typically used to express positive specifications, and we can use the same approach we did in regular expressions. If, with each event we observe, we keep track of the LTL formula that still needs to be satisfied using the derivative function, it suffices to check whether there exist traces that would match with the remaining derivative formula. If we manage to define such a function *cannotMatch*(e), we can implement monitoring in exactly the same way as we implemented it for regular expressions.

Let us consider, once again, the LTL property which states that greylisting a user should be followed immediately by downgrading the user to normal status:

```
globally (
  [after UserInfo.makeGreylistedUser()] implies
    next [before UserInfo.makeNormalUser()]
)
```

Let us assume that we observe a call to and return from the method `UserInfo.makeGreylistedUser()`, immediately followed by a call to the method `UserInfo.openSession()`. We would apply the derivative function on these events one-by-one as they are observed, starting with the property as the initial formula.

Applying the derivative on the initial formula and with event `before UserInfo.greylist()`, we would obtain the following new formula:

```
false implies [before UserInfo.makeNormalUser()] and
globally (
  [after UserInfo.makeGreylistedUser()] implies
    next [before UserInfo.makeNormalUser()]
)
```

If we now take the derivative of this formula with respect to `after UserInfo.makeGreylistedUser()` (the second event observed), we would obtain:

```
false implies false and
true implies [before UserInfo.makeNormalUser()] and
globally (
  [after UserInfo.makeGreylistedUser()] implies
    next [before UserInfo.makeNormalUser()]
)
```

Finally, we take the derivative with respect to the third event `before UserInfo.openSession()`, obtaining:

```
false implies false and
true implies false and
false implies [before UserInfo.makeNormalUser()] and
globally (
  [after UserInfo.makeGreylistedUser()] implies
    next [before UserInfo.makeNormalUser()]
)
```

If we look at the last formula obtained, we can see that the second conjunct equates to `false` and therefore so does the whole formula. This means that no future events can ever lead to the satisfaction of the formula, and we can thus report a violation.

All that remains to be done is thus to define the *cannotMatch*() to check whether an LTL formula is satisfiable. The problem is that an exact solution is known to be of high computational complexity. In order to reduce the overheads in runtime verification we can use a simpler function which identifies immediate unsatisfiability without checking for contradictions into the future i.e. the function would identify the unsatisfiability of `false and [a]` but not of `not [a] and [a]`. In the latter case, unsatisfiability would be captured once another event is observed.

This weak version of *cannotMatch*(e) is implemented as method `Boolean cannotMatch()` in the classes that inherit from the `LTL` class. Using this, we can implement the monitoring of LTL formula in much the same way we implemented it for positive regular expressions: for each LTL formula (i) we keep a variable representing the current formula to match; (ii) we update it using the derivative function with every relevant event received; and (iii) report a violation if the formula *cannotMatch*() is ever satisfied. The translation to EGCL is practically identical to that used for regular expressions.

Exercise 9.6

By referring to how positive regular expressions are monitored, implement the following for LTL scripts:

1. Create a method `String toEGCL()` in the `LTLScript` and `LTL` class, which will generate an EGCL script object for a single LTL property. The EGCL should handle calls to the derivative function and checks for unsatisfiability as already explained for positive regular expressions.
2. Modify `Main.java` to use this method to generate an AspectJ file and a verification Java class file.
3. Extend the solutions to work for multiple LTL properties in the script. Modify `Main.java` accordingly.

9.4 Parameterised LTL Formulae

As we have already noted, many properties should be applied on a per-object basis, so as to monitor the individual instances of a class independently. Take the example we saw earlier — the property which says that greylisting a user should be followed immediately by giving them a *normal user* status. We expressed this in LTL (using our scripting language) as follows:

```
property "Greylisting must be followed by changing status to normal user"
  good behaviour {
  globally (
    [after UserInfo.makeGreylistedUser()] implies
      next [before UserInfo.makeNormalUser()]
  )
}
```

However, the property does not ensure that `makeNormalUser()` is applied to the same object as `UserInfo.makeGreylistedUser()` was applied to just before. We would like to ensure that a separate monitor is enacted for each instance of `UserInfo` to monitor the objects separately. Adopting syntax similar to that used in parameterised regular expressions, we will specify such a parameterised LTL formula as follows:

```
property "Greylisting must be followed by changing status to normal user"
  foreach target UserInfo
  good behaviour {
  globally (
    [after UserInfo.makeGreylistedUser()] implies
      next [before UserInfo.makeNormalUser()]
  )
}
```

As in the case of parameterised regular expressions we do not allow the adding of new monitoring state (using the `keep` keyword) since, just like regular expressions, LTL formulae do not support conditions or actions. As in the previous examples of parameterised properties, the methods monitored in a parameterised property must target the same class.

Exercise 9.7

Write parameterised LTL specifications for the following properties:

- Property 5: *"Once a user is disabled, he or she may not withdraw from an account until the administrator enables them again"*.
- Property 10: *"Logging can only be made to an active session (i.e. between a login and a logout)"*.

chapter09-parameterised

Compilation of parameterised LTL formulae follows the same lines as parameterised regular expressions, generating parameterised EGCL, keeping a record of the current LTL formula in the EGCL replicated monitoring state.

To monitor parameterised LTL formulae, we will produce a parameterised EGCL script object and transform it into AspectJ. The support code for handling parameterised LTL can be found in chapter09-parameterised under the folder `MyRVTool`. New code of interest includes:

- `LTLScript.java` provides a new method to access the list of parameterised LTL formulae given in the script, called `ArrayList<ParameterisedLTL> getParameterisedLTLFormulae()`.
- `ParameterisedLTL.java` represents parameterised LTL formulae providing the following getters: (i) `String getTargetType()` returns the type of the target; and (ii) `LTL getLTL()` returns the LTL formula.

Exercise 9.8

Review the code provided and the approach taken to monitor parameterised regular expressions and implement the following:

1. Modify the code you wrote earlier to monitor LTL formulae so as to handle parameterised LTL formulae.
2. Change `Main.java` to read an LTL script file (which may use parameterisation) and output an AspectJ file and the verification class Java file.
3. Test the tool using the properties from the previous exercise.

chapter09-parameterised/MyRVTool

9.5 Conclusions

In this chapter we have seen how a logic with explicit temporal operators can be used for runtime verification. Having explicit temporal operators typically makes the expression of many temporal properties more straightforward. The complexity of the underlying verification code has, however, increased substantially. On the other hand, from a user's perspective, the challenge remains solely that of expressing a specification using the new logic.

It is worth mentioning that the notion of time in LTL (as with regular expressions) is rather limited. Essentially it has a notion of discrete events happening in order, but does not provide any means of referring to actual time, thus limiting its expressivity. Certain properties, such as Property 11: "*A session should not be opened in the first ten seconds immediately after system initialisation*", cannot be expressed in LTL without having access to actual timestamps. In the next chapter we will see how we can deal with real-time properties in our tool.

Chapter 10
Monitoring Real-Time Properties

Many of the properties we have been using in the book are temporal in the sense that they dictate constraints on the temporal order in which events should occur, e.g. *"The transaction system must be initialised before any user logs in"*.While these kind of properties are highly prevalent in software, we sometimes need to deal with real-time constraints, in which the time when an event happens is important, such as: *"A session should not be opened in the first ten seconds immediately after system initialisation"*or *"A new account must be approved or rejected by an administrator within 24 hours of its creation"*.

Real-time properties[1] occur frequently even in systems which are not typically associated with real-time applications. For example, in the context of graphical user interfaces (be it a website or a desktop application) one would be interested in making sure that users receive a response to their requests within a particular time period. In the context of user account management, one might be interested in ensuring that an activation is performed within fifteen minutes since the verification email has been sent, or that the account is deactivated if not used for longer than six months.

This chapter will explain how the specification fragments considered so far can be extended to allow the expression and monitoring of real-time properties. There are also a number of particular implications which need to be taken into consideration when monitoring such properties since doing so may modify the timing of a system, typically by slowing it down. These issues are discussed in the last section of the chapter.

---

[1] Perhaps more appropriately the real-time properties considered in this book should be called *soft* real-time properties as opposed to hard real-time properties where the granularity of the time bounds is typically very small and the deadlines are strict. In the context of Java where the garbage collector can interrupt normal execution at any time, hard real-time properties are virtually impossible to check for. For readers interested in this area, a number of JVM implementations have been created specifically for this purpose and a *Real-Time Specification for Java* has been drawn up to guide such implementations. However, in this book we choose to stick to systems using the mainstream JVM.

C. Colombo, G. J. Pace, *Runtime Verification*,
https://doi.org/10.1007/978-3-031-09268-8_10

10.1 Monitoring Real-Time Properties

Without any modification to our tool, we can already support real-time property monitoring. Consider the Property 11: *"A session should not be opened in the first ten seconds immediately after system initialisation"*.By using system calls to get the time of initialisation, storing it, and then measuring the elapsed time every time there is an attempt to open a new session, we can identify violations:

```
BackEnd.initialise(..) |
  -> { Verification.initialisedTime = System.currentTimeMillis(); }

UserInfo.openSession(..)
  | System.currentTimeMillis() -
      Verification.initialisedTime < 10*1000
  -> { Assertion.alert("P11 violated"); }
```

By storing the time of initialisation in a field in the `Verification` class, we can check whether opening a session came too soon. This works well, although running tests for the property would require us to wait for 10 minutes to check certain scenarios. We encapsulate the system time call within a new class `TimerManager`, also providing a `currentTimeMillis()` method to get the current timestamp (measured in milliseconds but additionally allowing for the fastforwarding of time. This functionality is provided through the method `fastforward` which takes as input the number of milliseconds the clock needs to be taken forward, and which is used in the `Scenarios` class to simulate the passage of time:

```
BackEnd.initialise(..) |
  -> {
    Verification.initialisedTime = TimerManager.currentTimeMillis();
  }

UserInfo.openSession(..)
  | TimerManager.currentTimeMillis() -
      Verification.initialisedTime < 10*1000
  -> { Assertion.alert("P11 violated"); }
```

Encoding timing constraints manually, is feasible for simple properties. But now consider Property 15:*"A new account must be approved or rejected by an administrator within 24 hours of its creation"*. Once a request for a new account is received, if nothing happens within 24 hours, the monitor should flag the violation. Here we have a problem: the violation does not coincide with any particular system event, but rather is due to the passing of time (exactly one day after the request to open an account). We note that unlike the former property which sets an lowerbound to the time of triggering of an event, the latter sets an upperbound and would thus require additional machinery to detect violations thereof.

If there is no need to detect a violation the moment it occurs, we can wait until the next system event occurs (the first event happening after the deadline has elapsed) when the violation can be discovered and reported. An alternative solution is to have the monitor include a polling mechanism which artificially creates a heartbeat event, allowing for checks whether any timing violations have occurred with every heartbeat. Violation would be reported with a delay, but never by more than the time between two consecutive heartbeats. However, if a violation is to be reported as soon as it occurs, then a timer event needs to be set to trigger at the exact time of violation.

For Property 15, the request for a new bank account would set up a timer to fire after 24 hours. If an approval or rejection of the account is received in time, the timer would be stopped, but if neither happens within one day, the timer would fire, triggering the detection of the violation. We will introduce additional machinery to handle such timers as part of our runtime verification tool in the next section.

Exercise 10.1

Use EGCL and system time (`TimerManager.currentTimeMillis()`) to monitor the following real-time properties in FiTS:

- Property 12: *"Once a blacklisted user is whitelisted, they may not perform any single external transfer worth more than \$100 for 12 hours"*.
- Property 13: *"A user may not have more than three accounts created within any 24 hour period"*.

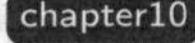

10.2 Timers

To support different types of real-time properties, such as ones with an upper-bound by when an event is to happen, we build a richer timer implementation to complement EGCL (and other formalisms that use it). Instead of just keeping track of the system time, we can create manipulate multiple timers each of which may fire an event after a user-specified duration elapses, much like a countdown clock. This is provided in the `Timer` class included in the code. We will be using these timers to handle real-time in the rest of this chapter.[2]

Consider Property 14that says that *"An administrator must reconcile accounts within 5 minutes of initialisation"*. Upon initialisation, a timer is created and set to fire in 5 minutes time. The monitor also keeps track of whether accounts have been reconciled, and when the timer eventually fires, a violation is reported if no such reconciliation took place:[3]

```
// Set up a 5 minute timer
Timer initialisationTimer = new Timer(5*60*1000l);
Boolean fitsReconciled = false;

after BackEnd.initialise(..) |
  -> {
    initialisationTimer.reset();
    fitsReconciled = false;
  }

after FrontEnd.ADMIN_reconcile(..) |
  -> { fitsReconciled = true; }

before Timer.fire(..) target(Timer t)
  |  !fitsReconciled
  -> { Assertion.alert("P14 violated"; }
```

Note that the `fire()` method in the *Timer* class is executed automatically when the timeout happens. The call will not do anything but serves as an event that can be intercepted (like other method call) using EGCL. The call to the `reset()` method upon intercepting system initialisation starts off the timer. To capture this property, we only have a single timer, meaning that reacting on the execution of any `fire` suffices. We will look into triggering on particular timers firing (when we have more than one) later on in this chapter.

Exercise 10.2

It is worth noting that the implementation of the runtime monitor for Property 14 assumes that a second initialisation taking place within the 5 minute window will cancel the requirement for an accounts reconciliation for the first initialisation and restart the timer. Modify the EGCL specification such that reconciliation is required within 5 minutes of an initialisation or before the next initialisation, whichever comes first. Re-

[2] Internally, the `Timer` uses the `TimerManager` class to obtain the current timestamp, thus automatically supporting the fastforwarding of timers.

[3] For better modularity, the timer and Boolean flag should be in the `Verification` class. We just show the variable declarations here to keep the presentation simple.

organise your code to have all verification code handling system events in a `Verification` class.

Use the *Timer* implementation provided to monitor and verify the following properties:

- Property 15: *"A new account must be approved or rejected by an administrator within 24 hours of its creation"*.
- Property 16: *"A session is always closed within 15 minutes of user inactivity"*.

chapter10

Thus far we have shown the basic use of the `Timer` class. In the following section, we elaborate on other advanced features which are provided and which help in more complex specifications.

10.3 Advanced Timer Features

The `Timer` class provides other functionalities useful for monitoring and verification.

Enabling and disabling a timer: Enabling or disabling the firing of a timer can be useful to turn off timers which are no longer required beyond a particular point in the system's lifetime. The `Timer` class provides this functionality through the `disable()` and `enable()` methods. It is worth noting that the timer continues running when disabled, and it is just the firing that is disabled.

Exercise 10.3

Use the `disable()` method in the `Timer` implementation provided to do away with the Boolean variable we used when monitoring Property 14:*"An administrator must reconcile accounts within 5 minutes of initialisation"*.

chapter10

Querying the timer value: The `Timer` class provides the method `time()` to query how much time has elapsed since the last reset of the timer (returning a `Long`, representing the time elapsed in milliseconds). For example, this can be used, for example, with lowerbound properties to check whether enough time has passed since a particular event. Note that a timer continues running even after it fires an event.

Exercise 10.4

Modify the monitoring code provided in Section 10.1 monitoring Property 11(*"A session should not be opened in the first ten seconds immediately after system initialisation"*) to use the `time` method provided in the *Timer* instead of referencing the system time.

chapter10

Labelling the timer: If there are multiple timers, it is crucial to distinguish between the timers by assigning each a name to allow for checking which timer has fired. A `Timer` constructor which takes a name and the timeout period is provided: `Timer(String id, Long delay)`. The identifier can be accessed using `getIdentifier()` method.

Consider extending Property 14(*"An administrator must reconcile accounts within 5 minutes of initialisation"*) with the additional constraint that the system cannot be initialised more frequently than once in any 15 minute time window. In this case, we would need two timers to handle the combined properties:

```
// Set up two timers: a 5 minute one for reconciliation and a 15 minute
// one for initialisation
Timer reconciliationTimer =
  new Timer("ReconcilationTimer", 5*60*1000l);
Timer initialisationTimer =
  new Timer("InitialisationTimer", 15*60*1000l);

Boolean initialisationAllowed = true;

*.initialise(..)
  |  initialisationAllowed
  -> {
    reconciliationTimer.reset();
    initialisationTimer.reset();
    initialisationAllowed = false;
  }

*.initialise(..)
  |  !initialisationAllowed
  -> {
    Assertion.alert(
      "Initialisation attempted before 15 minutes elapsed"
    );
  }
...
Timer.fire(..) target (Timer t)
  |  t.getIdentifier().equals("InitialisationTimer")
  -> { initialisationAllowed = true; }

Timer.fire(..) target (Timer t)
  |  t.getIdentifier().equals("ReconcilationTimer")
  -> {
    Assertion.alert("P14 violated");
  }
```

Note how the name of the timer is used to identify which timer has fired in order to take appropriate action.

Exercise 10.5

Complete the code from the example above and then modify the code to make do without the `initialisationAllowed` variable.

Add rules to monitor for the property that *"Initialisation should not be performed within 30 seconds of reconciliation"*.

chapter10

Timers which never fire: It is sometimes useful to set up a timer to keep track of time without a firing timeout. For example, in the previous example, we could have used such a timer to keep track of time since the last initialisation (to be accessed only through the `time()` method. Such timers can be created leaving out the timeout value in the constructor e.g. `new Timer("LoggedInTime")` or even simply `new Timer()` if the timer need not be named. It is worth keeping in mind that such timers will only start running once reset.

Pausing and resuming the timer: In certain cases, we will need to pause and resume a timer. For instance, we might want to keep track of total time a user spent logged in, or we might want to exclude any time in a state. The `Timer` class includes methods `pause()` and `resume()` to perform these actions on a timer.

Exercise 10.6

Recall Property 12:*"Once a blacklisted user is whitelisted, they may not perform any single external transfer worth more than $100 for 12 hours"*. Modify the property and its implementation such that if the user is greylisted during the 12 hour transfer-limit window, the time spent greylisted does not count towards that those 12 hours.

chapter10

The features described in this section should enable you to handle most real-time properties encountered in typical software systems. However, it is worth keeping in mind that there are numerous subtle issues which arise when dealing with such properties. We discuss some of these in the next section.

10.4 Real-Time Issues

Monitoring can have an effect on the underlying system, and this is particularly observable when it comes to real-time properties. The computational power required to monitor and verify a system (see Section 13.1 on measuring runtime verification overheads) will typically change the timing of that system[4] This situation calls for caution when monitoring real-time proper-

[4] This is not typically the case with non-real-time properties, since the order of events usually remains unchanged when introducing monitoring code. Still, there are situations in which race-conditions may be introduced or resolved because of the overheads due to monitoring. Furthermore, if the system's logic is such that it may act differently depending on runtime measures of time elapsed, memory used, number of active threads, etc., introducing monitoring can also change its behaviour.

ties since some properties may be satisfied by the system but violated the moment the monitor is introduced or vice-versa.

Apart from this so-called *observer effect*, there are a number of other issues to be taken into consideration when monitoring real-time properties:

Time never stops: Whenever a monitor transition triggers due to a timed event or using a timer-based condition, one should keep in mind that the execution of the transition itself is consuming time. Therefore even if a guarded command has a timer check in the condition to make sure that no more than 10 seconds have elapsed, it may be the case that the condition is satisfied, but by the time the action is executed and control is given back to the system, more than 10 seconds would have passed.

Scalability and concurrency issues: Recall from the previous section that timer events are events which are not triggered by the target system, and is spawned on a separate thread. Unfortunately threads may cause concurrency issues, and are relatively expensive to spawn in Java. Using a naïve approach where each timer is a separate thread would be impractical for any system with a substantial number of timers. A solution would be to share a single thread between all the timers, taking care to handle the concurrency issues which might arise, which is how the `Timer` library is implemented.

Unbounded number of timers: Consider the property: *"Sessions may not be open for longer than 24 hours"*. With every opening of a new session, we can start a new timer to ensure that it is closed before 24 hours have elapsed. Unless the number of concurrent sessions is capped by the system, the number of timers we will need is not statically bounded, which would cause the monitoring footprint to grow substantially at runtime. This can be mitigated by having a single timer but with multiple firing times — you can look at the implementation of the `addFiringTime` method in the `Timer` class which allows precisely this.

While this introduction to real-time property monitoring is not meant to be exhaustive, we have introduced means of monitoring of soft real-time properties which are typically very useful in real-world systems.

10.5 Conclusions

Real-time properties are common in many real-life systems, but challenging (and intricate) to monitor using plain assertions. This combination makes real-time properties an ideal candidate for benefiting from runtime verification techniques. Checking temporal lowerbound properties can be done with the infrastructure presented up till the previous chapter, but properties such as temporal upperbound ones greatly benefit from the use of a richer timer implementation to allow for a reduction in monitoring code complexity. If we want to take the approach further, we can even adopt a logic that inherently supports real time, thus abstracting away from manual instrumentation of timers in the properties. Such logics are beyond the scope of this book, but pointers can be found in Appendix A.

Chapter 11
Reactive Runtime Monitoring

Until now, we have focused on the detection of unexpected software behaviour without going into what to do if such behaviour is detected, limiting ourselves just to the logging of such violations. Keeping record of observed violations can be seen as a basic form of a verification reaction, providing information to developers, notifying them of the existence of errors in their code or misuse of the software system. However, without any further intrusion into the system, these benefits are limited to addressing issues long after the system has continued. In this sense, if we limit ourselves to the logging of violations, then we might as well have the system dump a log of relevant events and have runtime monitoring take place on that log of events, thus not adding overheads to the system[1]. Runtime verification, however, provides us with an opportunity to act algorithmically in case of violations in an online and immediate manner, thus mitigating the propagation of problems as soon as they arise.

This chapter will take the reader through a number of strategies which can be employed to have runtime verification-triggered actions firing as a reaction to observations of the monitored system: starting from actions which attempt to correct errors discovered to techniques which attempt to detect upcoming violations and stop them before they even occur. For instance, if the monitor realises that a user has just carried out a transaction which they should not have been allowed to carry out, reparation code can reverse the transaction or, if that is not possible, disable the user temporarily to avoid further such transactions. On the other hand, if we have sufficient information to decide that a transaction should not be carried out *before* it takes place, then we can just stop it from happening. We discuss these and other solutions to reparation in this chapter.

Although in this book we have largely limited our view of runtime monitoring as a means of identifying system errors, nothing stops us from taking the approach one step further and use it to identify points-of-interest during

[1] We will, in fact, look at this alternative architecture in the next chapter.

C. Colombo, G. J. Pace, *Runtime Verification*,
https://doi.org/10.1007/978-3-031-09268-8_11

the execution of the system, and inject additional functionality accordingly. For instance, a monitor can be used to identify users of FiTS that regularly transact above a certain value per day, and trigger a reaction to automatically upgrade them to gold users. This use of runtime monitoring is typically referred to as *monitor-oriented programming,* in which specifications identify points-of-interest (not violations), and trigger reactions intended to add functionality rather than carry out reparation logic.

In all these cases, the monitors are interacting with the system and modifying its behaviour. This comes at a risk, and we will also briefly discuss how systems can be designed to be aware of monitoring intrusion and take this into account during development.

Note that, in order to discuss this issue, we will be limiting ourselves to examples using EGCL specifications to keep the examples simpler. However, the techniques apply to any specification language, since what we really care about is reacting at points-of-interest (violations or otherwise) no matter how they are identified.

11.1 Runtime-Verification-Triggered Reparations

We will start by looking look at ways of handling *reparations* — code to be executed to repair the system after a violation has been discovered. Recall Property 1:*"Only users based in Argentina can be gold users"*. The first thing to do is to identify what reparatory action we would like to trigger if a violation is discovered. This is dependent on the system and the severity of the violation. Should we simply notify the system administrators that the property has been violated and let them handle it manually? Should we take away the gold status from the user? In that case, what should the new status be? Or should we take more drastic action, disabling the user and charging them an administrative fee?

For the sake of the example, we choose to give the user a normal user status in case of a violation. The simplest way of triggering the reparatory action is to include it as part of the EGCL rules which identify the violation:

```
after UserInfo.makeGoldUser() target (UserInfo u)
  |  !u.getCountry().equals("Argentina")
  -> {
    // Log violation
    Assertion.alert("P1 violated");
    // Reparation
    u.makeNormalUser();
  }
```

The problem with this approach is twofold. Firstly, we are interleaving system-centric behaviour within the specification. Clearly, such reparation code should be written by the development team, not the quality assurance one, but the separation between the two responsibilities is now lost. This can be resolved by having the development team provide the reparation code for property violations themselves, for instance in a separate static class with methods to execute the reparation for each property e.g. `Reparation.trigger(Integer propertyNumber)`. The rule would only be changed slightly, but would keep apart the verification from the system logic:

```
after UserInfo.makeGoldUser() target (UserInfo u)
  |  !u.getCountry().equals("Argentina")
  -> {
    // Log violation
    Assertion.alert("P1 violated");
    // Reparation
    Reparation.trigger(1);
  }
```

However, there is a second problem with the original solution that is still not resolved: we are directly changing the system state from the specification. This can cause problems in the system since the intrusion may not have been taken into account by the developers. In this case, for instance, there may be an update of the user status without a corresponding update to the user interface. The system may be updating the user interface based on calls to the `FrontEnd` class methods, and may be showing that the user is now a gold user despite the fact that their status was reverted back to normal.[2] Having the reparations fall under the responsibility of the development team goes a long way in addressing this issue, but still, the reparation code is an independent block of code which can be triggered at various points during the execution of the system. Developers working on code maintenance of the main system code in the future may not be aware of the reparation code that can be triggered by a violation, resulting in new problems arising.

One solution is to use the exception handling mechanism in Java, and have the runtime verifier trigger an exception if a violation is discovered. This can be handled in an extended version of EGCL with a rule as follows:

[2] Perhaps this would indicate that we should have called the respective user interface level method as a reparation, but this may still be problematic since, for instance, calls to these methods may be assumed to be actual user requests which would not be the case here.

```
after UserInfo.makeGoldUser() target (UserInfo u)
  throws RVExceptionProperty1
  |  !u.getCountry().equals("Argentina")
  -> {
    // Log violation
    Assertion.alert("P1 violated");
    // Reparation
    throw new RVExceptionProperty1("...");
  }
```

By specifying and throwing an exception specific to the property being violated,[3] we can explicitly throw the exception to be caught by the system and handled using a `try-catch` block. In order to do so, we would have to enclose calls to `makeGoldUser` with an appropriate exception handler:

```
try {
  ts.getUserInfo(uid).makeGoldUser();
} catch (RVExceptionProperty1 rvX) {
  // Reparation
  ts.getUserInfo(uid).makeNormalUser();
}
```

This approach comes with a number of advantages. Firstly, the system code is written in a manner that explicitly refers to potential runtime failure, and includes appropriate reparation code. Furthermore, the developers can choose at which level to handle such runtime errors — the exception could have been propagated further up and handled at the user interface level, for instance.

One downside of this approach is that the developers must be aware of where violations of the properties may occur and their respective exceptions raised.

Another downside of this approach becomes apparent when we have events that trigger before, rather than after a method call. If the EGCL rule were to trigger on `before UserInfo.makeGoldUser()` events and throw an exception, the exception would effectively bypass the execution of the `makeGoldUser()` method altogether. Although this may be useful in some cases (we will discuss this later), we may be eliminating certain program logic which we may want to keep. Furthermore, the developer has to be aware of the modality of the event being handled, which is not documented anywhere except in the specification script itself.

Finally, some system code may have to be re-engineered to have it work as expected. Consider using this exception-based approach to monitor Property 2:*"The transaction system must be initialised before any user logs in"*.

[3] The additional EGCL syntax to declare that the rule may throw this particular exception is required due to the way AspectJ and Java handle exception declaration.

This may cause an exception to be raised whenever openSession() is called before initialisation. However, when we look at the way openSession() is used in the rest of the code, we find that it returns the new session id, or -1 in case of failure to open a session as follows:

```
if (u.isEnabled()) {
  return (u.openSession());
} else {
  return -1;
}
```

If order to handle the exception around the line using openSession(), we would have to ensure that the reparation carries out alternative logic, still returning -1 to indicate failure:

```
if (u.isEnabled()) {
  try {
    return (u.openSession());
  } catch (RVExceptionProperty2 rvX) {
    // Reparation
    return -1;
  }
} else {
    return -1;
}
```

As these examples show, programming reparations is not an easy task, particularly since they lie in a grey area between the system-under-scrutiny and the specification. Although they impact the behaviour of the former, they are triggered independently by the latter. We have presented different ways in which such code can be organised, and it typically boils down to how the system is designed and built. However, what is clear is that the earlier that runtime verification is planned as an inherent part of the system to be deployed, the better. One of the most challenging aspects of adding runtime verification to a system already developed and deployed is that of introducing ways of dealing with failure.

Exercise 11.1

The reparation we used for Property 1 (*"Only users based in Argentina can be gold users"*) was to downgrade the user to normal status. However, the user may have had silver status before the attempt to give them gold status. Modify the reparation instrumentation rules to ensure that in case of a violation, the user is restored to their previously held status, and not blindly to normal.

Review the EGCL specifications developed in Chapter 6 for the properties listed below. Implement and add reparations as you deem fit for these properties. Instrument the updated specifications and test them out using the code for the runtime verification tool provided.

- Property 3: *"No account may end up with a negative balance after being accessed"*.
- Property 4: *"A bank account approved by the administrator may not have the same account number as any other bank account already existing in the system"*.
- Property 5: *"Once a user is disabled, he or she may not withdraw from an account until the administrator enables them again"*.
- Property 6: *"Once greylisted, a user must perform at least three incoming transfers before being whitelisted"*.
- Property 7: *"No user may request more than 10 new accounts in a single session"*.
- Property 8: *"The administrator must reconcile accounts every 1000 attempted outgoing external money transfers or an aggregate total of one million dollars in attempted outgoing external transfers (attempted transfers include transfers requested which never took place due to lack of funds)"*.
- Property 9: *"A user may not have more than 3 active sessions at any point in time"*.
- Property 10: *"Logging can only be made to an active session (i.e. between a login and a logout)"*.

chapter11-reparations

11.2 Runtime Enforcement

In the previous section, we looked at reparations — ways of making up for the fact that a property has already been violated. However, in some cases we might realise that a property will be violated before it happens. *Runtime enforcement* takes this approach, identifying when a property is about to be violated, and stepping in to avoid the violation happening in the first place.

Let us consider once again Property 1:*"Only users based in Argentina can be gold users"*. We captured violations of the property at the moment a user has just been given gold status using the `after UserInfo.makeGoldUser()` event. However, by inspecting the code, we know that a user will not change his or her residence address *during* the execution of `makeGoldUser()`, which means that we could have captured the occurrence just before the call to the method instead:

```
before UserInfo.makeGoldUser() target (UserInfo u)
  throws RVExceptionProperty1
  |  !u.getCountry().equals("Argentina")
  -> {
    // Log enforcement
    Assertion.alert("P1 enforced");
    // Notify system of upcoming failure
    throw new RVExceptionProperty1("...");
  }
```

Note that at this stage, the status of the user has not yet been updated, and the exception stops the method from being executed, avoided a violation, and effectively *enforcing* the property. Note that we still log the enforcement activity since developers should be aware of it, in order to identify and fix the bug which would have led to such a violation. The exception handling code in the system can remain in place except that no reparation code is necessary:

```
try {
  ts.getUserInfo(uid).makeGoldUser();
} catch (RVException1 rvX) {
  // Suppression of makeGoldUser() for enforcement of P1
};
```

In this manner, by *suppressing* execution we will have ensured that the system has never violated the property in the first place. The caveats and concerns discussed earlier still apply. Other code e.g. user interface code, may be assuming that the action of updating the user's status to gold succeeded, and may result in a mismatch between the internal state and that of the user interface. Furthermore, we assume that the last action which led to the violation is the one that ought to be suppressed. If we had a method for users to update their residence address, we would need to add an additional rule to check that changing the residence of a gold user to a non-Argentinian address should trigger a violation. If we apply an enforcement approach, we would disallow the user from changing their address when, in fact, reparation might be more appropriate in this case, allowing the change of address but downgrading the user status.

Exercise 11.2

Update the reparations used in the previous exercise in order to enforce the properties by suppressing behaviour (rather than performing a reparation):

- Property 3: *"No account may end up with a negative balance after being accessed".*

- Property 4: *"A bank account approved by the administrator may not have the same account number as any other bank account already existing in the system"*.
- Property 5: *"Once a user is disabled, he or she may not withdraw from an account until the administrator enables them again"*.
- Property 6: *"Once greylisted, a user must perform at least three incoming transfers before being whitelisted"*.
- Property 7: *"No user may request more than 10 new accounts in a single session"*.
- Property 9: *"A user may not have more than 3 active sessions at any point in time"*.

chapter11-enforcement

Certain properties cannot be enforced simply by suppressing events. Properties which lead to a violation because of the lack of an event rather than its presence require the insertion of new actions or the replacement of existing ones in order to enforce correct behaviour. Let us consider Property 2:*"The transaction system must be initialised before any user logs in"*. If the runtime monitor observes a request to open a new session before FiTS is initialised it has two options. The first is to suppress the `openSession()` method call. The second is to have the monitor initialise FiTS in order to make the call to open a session a legitimate one:

```
if (u.isEnabled()) {
  try {
    return (u.openSession());
  } catch (RVException2 rvX) {
    // Enforcement
    ts.initialise();
    return(u.openSession());
  }
} else {
    return -1;
}
```

Needless to say, such insertion of new behaviour once again comes with its caveats. If FiTS has been designed such that initialisation is expected to be performed by an administrator when they want the transaction system to go live, such an enforcement is a dangerous one, since it provides any user of the system the ability to force FiTS to go online.

Exercise 11.3

Specify runtime enforcers of the properties by insertion of new behaviour:

- Property 8: *"The administrator must reconcile accounts every 1000 attempted outgoing external money transfers or an aggregate total of one million dollars in attempted outgoing external transfers (attempted transfers include transfers requested which never took place due to lack of funds)"*.
- Property 10: *"Logging can only be made to an active session (i.e. between a login and a logout)"*.

chapter11-enforcement

Another place where enforcement through insertion can be useful is in the case of deadline properties — properties which state that something must happen within a certain amount of time. Recall Property 16:*"A session is always closed within 15 minutes of user inactivity"*. In this case, we can trigger an action to close the active session either as soon as 15 minutes have elapsed (thus fixing the problem as soon as it occurs) or just before 15 minutes have elapsed (thus ensuring that the property is never violated in the first place).

We can build an enforcer for this property by creating a timer for each new user session opened.[4] We can keep track of these timers indexed by the user session in the `Verification` class:

```
public class Verification {
  static public HashMap<UserSession, Timer> timers =
    new HashMap<UserSession, Timer>;
  . . .
}
```

We can now add EGCL rules to create a new timer every time a new session is opened, reset the timer every time a user activity is detected[5] and reset every time the session is closed:

[4] Note that ideally we would have used the `foreach` construct to manage the timers. However, this is not possible (with the minimal infrastructure built for this book) as further on we add the timer event whose target is not part of the `UserSession` object.

[5] For simplicity we just consider user logging as relevant activity, but one can add other events such as account transfers if we wanted to.

```
before UserSession.openSession() target(session) -> {
  // In case a session was already opened
  Verification.timers.remove(session);

  // Get new timer and add it to hashmap
  Timer timer = new Timer(
    "TIMEOUT PROPERTY 16," +
    session.getOwner() + "," + session.getId(),
    15*60*1000l - 500l
  );
  Verification.timers.put(session, timer);
  timer.reset();
}

before UserSession.log(..) target(UserSession session) -> {
  if (Verification.timers.containsKey(session) {
    Verification.timers.get(session).reset();
  }
}

before UserSession.closeSession() target(UserSession session) -> {
  if (Verification.timers.containsKey(session) {
    Verification.timers.get(session).disable();
  }
}
```

Note that we chose the timer to trigger half a second before the 15 minutes elapsed, but this can be changed depending on how developers decide to handle such a timeout. Finally we add a rule to handle timeouts and forcing the closure of the user session:[6] It is worth noting that in order for the timer to be able to close the session, one would need access to the FiTS FrontEnd object, which would have to be recorded in the Verification object (e.g. after initialisation, or whenever a session is opened through the front end) for later access.

```
before Timer.fire(..)) target (Timer timer) -> {
  String[] values = timer.getIdentifier.split(",");
  if (values[0].equals("TIMEOUT PROPERTY 16")) {
    // Obtain uid and sid from timer identifier
    Integer uid = Integer.parseInt(values[1]);
    Integer sid = Integer.parseInt(values[2]);

    // Close session to enforce property
    Verification.frontend.ADMIN_closeSession(uid, sid);
}
```

[6] The FiTS interface does not have a ADMIN_closeSession(..) method, but this can easily be added to allow the session closure to be registered as an administrator-enforced one.

Enforcing such timeouts is possible, though it also comes with its downsides — the intricate nature of the code being one of them, apart from the already discussed potential of undesirable interaction with existing code.

Exercise 11.4

Use timers to enforce Property 14:*"An administrator must reconcile accounts within 5 minutes of initialisation"*, by triggering the reconciliation process just before the timeout triggers.

Property 15:*"A new account must be approved or rejected by an administrator within 24 hours of its creation"*, is a timeout property which does not lend itself easily to runtime enforcement. The decision whether or not to approve an account should presumably be taken by the administrator. However, if such a property arises from a contractual agreement with the user, one may decide that any pending accounts should be decided in an automatic manner (one way or another) just before the deadline in order to ensure conformance. Implement an enforcer to reject requested accounts just before the timeout occurs.

chapter11-enforcement

In this section, we have looked at event suppression and insertion in order to enforce properties. Other possible runtime enforcement techniques not discussed in this chapter include event replacement and parameter modification. Although enforcement looks good on paper, since it ensures that the properties are never violated, it is not easy to introduce this approach to an existing system without risking unintended effects. If this approach is to be adopted, it should be planned as an integral part of the underlying system, and even so, great care should be taken to ensure coherence of the system state is maintained.

11.3 State Rollback through Checkpointing

The idea behind reparation and enforcement is that there is a point in time (when an event or timeout occurs, for instance), at which we can identify a violation or the risk of a violation occurring, and by changing the system behaviour (also at that point in time), we can fix adverse effects or avoid altogether the violation from occurring. However, there are properties for which violations can only be detected at a time later than that of the actual violation. For instance, by the time we detect potentially fraudulent behaviour of a user (by analysing sequences of transactions they are performing) the offending party may already have performed transactions which we would rather not have happened. In such cases, the only option is to carry out retrospective reparation.

As we have seen, programming reparations or enforcers is not a straightforward task. It requires intimate knowledge of the system and each recovery must be custom made for the property in question. A more general approach is to have the system be able to undo earlier behaviour, thus allowing us to fix the consequences of past actions through what is called *backward recovery*. This approach can be seen to lie somewhere between reparation and enforcement since the monitor will be allowing events which, later on and in hindsight, we would rather that they did not happen in the first place. Once we realise this, we can restore the system state as though those events never happened, essentially making it a form of enforcement.

There are different ways of performing backward recovery. In this section we will look at one of these techniques — the use of *checkpointing*, which involves saving a snapshot of the system state so as to be able to revert back to it if something goes wrong at a later stage. Such saving and restoring of a whole system is typically too prohibitive, but for many properties we need only be concerned with a particular object or group of objects. For instance, we may only be concerned with saving and restoring the state of a single user or of a single account.

In order to illustrate the use of checkpointing, we will add new user functionality to FiTS. We will allow the user to trigger *compound transactions*, consisting of a sequence of internal transaction across accounts, all of which are owned by the user initiating the compound transaction, and which take place without any further user interaction.

Exercise 11.5

FiTS is updated such that the `FrontEnd` class now includes a method to handle compound transactions: `USER_compoundTransaction(Integer sid, CompoundTransaction ct)`. In addition, two additional files are included:

- `InternalTransaction.java` provides the `InternalTransaction` class which allows the representation of a transfer across two accounts owned by the same user. The class provides methods to access the parameters of the transfer: `BankAccount getSourceAccount()`, `BankAccount getDestinationAccount()`, `Double getAmount()` and `UserInfo getOwner()`.
  The method `void execute()` carries out the transfer (if possible).
- `CompoundTransaction.java` allows the representation of a compound transaction using the `CompoundTransaction` class. The class provides access to the sequence of transactions making up the compound transaction through the method `ArrayList<InternalTransaction> getInternalTransactions()`.

> Review the code provided and implement method `void execute()` in class `CompoundTransaction`, which carries out the transactions (left to right).
>
> chapter11-checkpointing

The property which we would like to hold is that throughout a compound transaction, at no time should the balance of any of the user's accounts involved in the transfers exceed $10,000. Violation of the property can be captured at the internal transaction level in EGCL as follows:

```
after InternalTransaction.execute() target (InternalTransaction t)
  throws RVExceptionPropertyCompoundTransaction
  |  t.getSourceAccount().getBalance() > 10000
  -> {
    // Log violation
    Assertion.alert("Maximum account balance property violated");
    // Throw exception
    throw new RVExceptionPropertyCompoundTransaction("...");
  }
```

Applying reparations at the internal transaction level is straightforward: we can simply reverse the transaction. However, at the compound transaction level, we may want to ensure that if such a transaction fails halfway through, then the whole compound transaction should fail, and the state should be restored as though none of the transactions were executed. One way of applying this reparation is by looping through the transactions already carried out but in reverse order, and apply them again — this time with the funds flowing in the opposite direction. This results in rather intricate reparation code.

Another way of solving the problem is through the use of checkpointing. Since compound transactions deal with a single user, one can save the user accounts and their balances when a compound transaction starts, and simply restore the system to that state should the property be violated halfway through the execution.

We must decide where to save the snapshot. One way is to save it in the `Verification` class, which handles verification logic, but this requires substantial system-level code being written in the specification scripts. Another is to provide the functionality at the transaction system level, but this can become quite unwieldy as one may require saving snapshots of different parts of the system for different properties. A third option is to build snapshot saving and restoring of user accounts at the `UserInfo` level, which works out to be cleaner by making such logic local to where it matters:

```
public class UserInfo {
  private ArrayList<BankAccount> bankAccountsSnapshot;
  ...

  public void saveBankAccountsSnapshot() {
    // Clone the accounts individually
    bankAccountsSnapshot = new ArrayList<BankAccount>();

    for (BankAccount a: accounts)
      bankAccountsSnapshot.add((BankAccount) a.clone());
  }

  public void restoreBankAccountsSnapshot() {
    // Restore accounts
    accounts = bankAccountsSnapshot;

    // Reset snapshot
    bankAccountsSnapshot = null;
  }

}
```

Note that for this to work, `BankAccount` must be made into a cloneable class:

```
public class BankAccount implements Cloneable {
  ...
  @Override
  protected BankAccount clone() {
    try {
      return (BankAccount) super.clone();
    } catch (CloneNotSupportedException ex) {
      return null;
    }
  }
 }
```

This allows us to save the state of the user's accounts before the compound transaction starts using the following rule:

```
before CompoundTransaction.execute(..) target (ct) |
  -> { ct.getOwner().saveBankAccountsSnapshot(); }
```

We can now update the code to handle a user request to execute a compound transaction at the `FrontEnd` level to capture the runtime verification exception and restore the user's account balances:

```
public Boolean USER_compoundTransaction(
  Integer sid, CompoundTransaction ct) {
  try {
    return ct.execute();
  } catch (RVExceptionPropertyCompoundTransaction rvX) {
      // Restore balances to their original values
      ct.getOwner().restoreBankAccountsSnapshot();
      return false;
  }
}
```

Checkpointing thus allows us to rewind time selectively, restoring the relevant parts of the system state when failure is detected late and we would like to undo previous actions which are now, in retrospect, undesirable.

Exercise 11.6

Implement the checkpoint-based solution described above, and enhance it in the following manner:

- The execution of a single internal transaction can also result in a violation of the property we implemented above. Ensure that such violations are captured, and implement a reparation to reverse the single transaction being executed. Since compound transactions are executed by iterating over the constituent internal transactions, both reparations for the individual transaction and the compound one will be triggered. Make sure that your code triggers only the appropriate reparation.
- Snapshots can be considerably memory intensive, and it is always good practice to allow the garbage collector to dispose of snapshots once we know that we will no longer use them. Modify your specification to ensure that once a compound transaction successfully terminates, the snapshot can be garbage collected (by pointing the snapshot variable to null).
- Extend the code to allow a compound transaction to have not only internal transactions but also outgoing ones to other users. Use checkpointing to implement appropriate reparations. Is checkpointing appropriate to be used for this property?

chapter11-checkpointing

11.4 State Rollback through Compensations

Checkpointing is not always practical due to the additional memory requirements induced. In the example we saw in the previous section, we saved the status of all the user's accounts when only a small number of them may have changed during the compound transaction. It would have been even worse had we supported compound transactions between accounts of different users, which would have required saving the balances of all the accounts in FiTS, effectively duplicating the memory footprint of the system.

One can adopt a more fine-grained approach, for instance, saving the state of an account only when it is about to be changed, and if it has not yet been observed to change in this compound transaction. If a violation is observed, all the saved accounts can then be restored. The downside is that this requires substantial additional logic and the opportunity to introduce new bugs. The simplicity of the save-upon-starting, restore-upon-violation pattern is broken.

This approach of keeping track of things depending on what is happening is, however, the idea behind another backward recovery strategy: *compensations*. In a compensation-based approach, once we start a computation which may fail, we will keep track of an inverse action (or compensation) for each and every forward action carried out. If a failure occurs halfway through such a compound computation, we simply execute these compensations in reverse order. If no violation is observed by the time the compound action terminates, we can simply discard the compensations.

For example, consider a compound transaction consisting of transferring \$1000 from account A to B, then \$500 from B to C, and finally \$200 from B to D. Let us assume that we execute this compound transaction with the initial account balances of A, B, C and D being \$3000, \$0, \$9800 and \$500 respectively. Note that after the first two transactions have taken place, we will be notified of a violation (since account C will have over \$10,000). Now, if our system defines the compensation of a transfer of an amount from one account to another to be the reverse transaction i.e. the same amount being transferred from the destination to the source account, at the moment of violation we would have kept track of two compensations: a transfer of \$1000 from account B to A, and a transfer of \$500 from account C to B. All we need to do to undo the initial part of the compound transaction that took place is that of running these compensations in reverse order: starting from the compensation transfer from C to B, followed by compensation transfer from B to A. Note that we are only keeping track of the changes, rather than the status of all the accounts that may be affected, which makes this much more effective than naïve checkpointing. Furthermore, the compensations need not be exact inverses of the forward computation. We could, for instance, have enforced an additional charge as part of each compensation.

In order to incorporate compensation-based recovery we have to be able to (i) initiate compensation recording e.g. when a compound transaction is about to start; (ii) add compensations as each individual component successfully terminates; (iii) trigger compensation execution if a violation is observed; and (iv) clear compensations when they are no longer needed e.g. when a compound transaction terminates successfully.

When it comes to compensation representation, in the case of the property imposing a $10,000 cap on accounts during a compound transaction, we already have the `InternalTransaction` class to keep track of single compensations. In order to store a chain of compensations which we may trigger, we can either (i) keep a `CompoundTransaction` compensation object, starting with an empty compound transaction, and adding every new compensation to the front of the list for when it is possibly triggered, or (ii) use a stack data structure such as the `Deque` class in Java. We will adopt the latter solution since for some properties we may not have an appropriate custom data structure at hand, thus making it a more general and practical solution. Below is a class handling internal transaction compensations:

```
public class Compensations {
  private Deque<InternalTransaction> compensations;

  ...
  public void clearCompensations() {
    compensations = new Deque<InternalTransaction>();
  }

  public void addCompensation(InternalTransaction t) {
    compensations.push(t);
  }

  public void triggerCompensations(FrontEnd frontend) {
    for (InternalTransaction t: compensations) {
      frontend.ADMIN_transferBetweenAccounts(
        t.getOwner().getId(),
        t.getSourceAccount().getAccountNumber(),
        t.getOwner().getId(),
        t.getDestinationAccount().getAccountNumber(),
        t.getAmount()
      );
    }
    clearCompensations();
  }
}
```

Note that we chose to call the compensation transfers not through the user interface call available to the user `USER_transferOwnAccounts(..)` method but through a new method which will be reserved for administrator-triggered transfers. This ensures that any logging is done appropriately, but also allows for modified behaviour e.g. adding a charge to the transfer.

Where the compensation mechanisms are to be stored is another important issue. Should they be seen as part of the system, or should they be a part of the runtime verification specification? The former has the advantage of being handled by the development team, which has more intimate knowledge of the system, whilst the latter may be preferable since triggering of the compensation code is the responsibility of the verifier. In the code we will show in this section, we adopt the former approach, storing the compensations object in the `CompoundTransaction` class, and triggering it in case of a violation exception being raised from exactly the same property we saw in the previous section:

```
public class CompoundTransaction {
  private Compensations compensations =
    new Compensations();
  ...

  public void execute(FrontEnd frontend, Integer sid) {
    compensations.clearCompensations();
    try {
      for (InternalTransaction t: getTransactions()) {
        t.execute(frontend, sid);

        InternalTransaction compensation =
          new InternalTransaction(
            t.getOwner(),
            t.getDestinationAccount(),
            t.getSourceAccount(),
            t.getAmount()
          );
        compensations.addCompensation(compensation);
      }
    } catch (RVExceptionPropertyCompoundTransaction rvX) {
      compensations.triggerCompensations();
    }
  }
}
```

This approach ties the compensations tightly to the system. In the exercises, we will also look at the alternative of keeping the compensations within the scope of the property. Which approach is more appropriate depends largely on the system itself.

The example we have used in this section to illustrate the use of compensations is a relatively simple one. The compensations are composed of a single linear chain of uniform components (an internal transaction). In more complex systems and properties, one can have more intricate compensations, with components possibly ranging over different types of computation (one may be a transfer, while another may be the blacklisting of a user) and the compensation chain may be triggered in different ways depending on what way the property is violated. However, the use of compensations gives us a framework and design pattern to use to organise such history-based reparations.

Exercise 11.7

Complete the code shown in this section to handle compensations stored on the system-side within the `CompoundTransaction` object.

Modify the compensation class to support two types of compensatory actions: transfers and administrator reporting of individual transfers. If a compound transaction leads to a violation (of the $10,000 cap), any internal transaction which took place as part of it should not only be reversed, but also reported to the administrator. Implement this by making enriching the compensations for such internal transfers.

The code we have shown in this section managed the compensations on the system-side. An alternative is to store the compensations in the `Verification` object, and have EGCL rules to (i) reset the compensation stack when a new compound transfer starts; (ii) push compensations onto it whenever an internal transfer which is part of a compound transfer is executed; (iii) trigger the compensation stack if the property is violated; and (iv) clear the compensation stack once a compound transfer terminates without a violation. Implement this alternative solution and compare it to the one we have shown in this section.

chapter11-compensations

11.5 Monitor-Oriented Programming

We have looked at different ways in which we can trigger different forms of reparatory action in case a property is violated. Such actions modify the system behaviour but ensure that we are doing all that is within our power to mitigate problems arising from the violation. However, the use of runtime monitoring to trigger additional system-logic can also be useful in the absence of problems. We can see our specifications not as properties, but as system behavioural patterns we are interested in, and have the actions add or modify behaviour of the underlying system when such patterns are observed. This approach is called *monitor-oriented programming.*[7]

We will illustrate how monitor-oriented programming can help in extending system behaviour by adding new logic to FiTS to implement a promotional campaign which will upgrade users to silver status once they have cumulatively deposited at least $10,000 to their accounts. We can implement this new feature using an EGCL script which (i) keeps track of the deposits balance of each user; (ii) upgrades the user to silver status if the threshold is

[7] You will also see it referred to as *monitoring-oriented programming* in the literature.

reached; and (iii) resets the running balance if the user has been downgraded back to normal status:

```
foreach target (UserInfo u)
  keep (Double depositsBalance defaultTo 0) {

  after UserInfo.depositTo(account, amount)
    |  u.isNormalUser()
    -> {
      depositsBalance = depositsBalance + amount;
      if (depositsBalance >= 10000) {
        u.makeSilverUser();
      }
    }

  before UserInfo.makeNormalUser()
    |  !u.isNormalUser()
    -> { depositsBalance = 0; }
}
```

In this case, the added functionality required no modification to FiTS, but in some cases, particularly when the new functionality would be accessible to the users, we may need to add new code to the system. Consider, for instance, if we want to introduce functionality such that users may declare who their spouse is so that the upgrade to silver status would also be applied to their spouse. In order to be able to add this functionality, we would need to have the system extended to allow users to declare who their spouse is, and means of accessing this information. Once that is done, the changes to the promotional campaign can be updated to use this information, we can modify the rule which handles user upgrades in the following manner:

```
after UserInfo.depositTo(account, amount)
  |  u.isNormalUser()
  -> {
    depositsBalance = depositsBalance + amount;
    if depositsBalance >= 10000) {
      u.makeSilverUser();
      if (u.hasSpouse() && u.getSpouse().isNormalUser()) {
        u.getSpouse().makeSilverUser();
      }
    }
  }
```

Monitor-oriented programming can thus be used to change the behaviour of a system in a manner which allows for the new logic to be programmed and documented independently. Despite its advantages, the downside is that interaction between different features can be difficult to evaluate, possibly leading to unexpected bugs. Having said this, although not strictly a form of runtime verification, monitor-oriented programming is an instance of runtime monitoring, and one worth being aware of when looking into how runtime monitoring and verification work.

Exercise 11.8

Change the promotional scheme such that a user gets silver status once they reach a deposit balance of at least $10,000 in the same calendar month. This upgraded status lasts for 31 days, e.g. if a user has exceeded the $10,000 in deposits on 12 January, they will get silver status until 11 February. With every new deposit in January, the end of the upgrade status is pushed further ahead e.g. if the user then deposits another $10 on 20 January, the upgrade will last until 19 February. Upon reaching that date, the user will be downgraded back to normal status unless they had already accumulated $10,000 deposits in February.

Add system functionality to (i) allow users to authenticate themselves using `USER_authenticate()` method in the `FrontEnd` class (for simplicity, assume that authentication always succeeds); (ii) keep track whether a user has authenticated themselves in the `UserInfo` class; and (iii) in the same class, change `openSession(..)` to work only if the user has already been authenticated, and `closeSession()` to deauthenticate the user. Use monitor-oriented programming to ensure that authentication is required again after every 10 user transactions or $1000 worth of withdrawals, whichever comes first.

chapter11-mop

11.6 Conclusions

While runtime verification can be limited to the flagging of property violations, we can take it a step further, triggering reactions when unexpected behaviour is observed. We have shown different ways of structuring such additional behaviour, with the common theme being that unless the system has been designed with this in mind, the risks of undesirable interference with the system can be unacceptably high. In such situations, it may be better to do just enough so as to stop the unexpected behaviour from propagating, but not attempt full recovery. Having said this, if feedback from the runtime verification component is planned to be part of the system-under-scrutiny, the techniques we have shown can be used to correct for different types of misbehaviour.

If we fear that any form of online runtime verification may interfere with our system, we can choose to go to the other extreme and limit ourselves to the logging of relevant events that the system is doing. Verification can then be performed separate from the system, thus reducing runtime overheads and risk of interference. This offline approach to runtime verification will be explored in the next chapter.

Chapter 12
Offline Runtime Verification

In the last chapter, we have considered one extreme in the spectrum of monitor intrusiveness — having the runtime verification tool trigger changes in the system. However, even if we avoid such system feedback, changes in the system behaviour may still be possible, whether it is timing changes which may lead to new race conditions, or whether it is due to monitors locking resources which the system may otherwise need. The fact that the monitoring and verification code shares and utilises resources which would have otherwise been exclusively available to the system carries a risk that may be deemed to be unacceptable in certain domains.

To minimise the appropriation of system resources, runtime verification can be made less intrusive. By decoupling the system and monitor execution such that the system does not work in lockstep with the monitor, i.e. the system does not wait for the monitor to return control at every step, ensures that the system is not unnecessarily held up by the verification effort. This is referred to as *asynchronous runtime verification.*

We can even go one step further and have the system just log events of interest, which are then consumed independently, possibly on another machine or at times when the system is facing a low load. This is referred to as *offline runtime verification*, using which the monitoring and verification code is completely decoupled from the main system.

These approaches, and variants thereof, can be particularly useful if the system-under-scrutiny is running on a low-resource platform e.g. an internet-of-things device, or when it faces an uneven load e.g. an online trading platform which may face an increased number of transactions at particular times such as when the value of important shares suddenly drops. Needless to say, such approaches limit the power of runtime verification in that by the time errors are identified it may be too late to act upon them and their consequences. The costs and benefits of decreased coupling between the system and the verification code are to be carefully weighed when making a choice on the ideal way of monitoring the system in question.

C. Colombo, G. J. Pace, *Runtime Verification*,
https://doi.org/10.1007/978-3-031-09268-8_12

In this chapter we will explore the development of an offline runtime verification tool, which keeps a log during execution and then replays it for the monitor to pass judgement independently. This will help give a better understanding of how decoupling between the system and monitor can be achieved.

12.1 Logging of Events

Unless events-of-interest are extracted at the virtual machine or hardware level, or constrained to the monitoring of externally visible events such as network communication, monitoring requires code to be instrumented as part of the system to log events. Whether the logging writes to a database, transmits events over a network, or otherwise, a degree of intrusion is unavoidable.

Capturing events-of-interest can be done either by manually adding code to the system to log events or using AOP, which we will discuss further in this section. Needless to say, if we attempt to log all the events from the system, the overheads can still be prohibitive, and it is thus crucial that only relevant events are logged. One way to achieve this is to write a separate pointcut for each relevant event, but we can also use jointpoint filtering which AspectJ supports:

```
execution(
  public * *(..)) &&
  if (pointsOfInterest.contains(thisJoinPoint.getSignature().getName())
)
```

Unlike `call` (which we used earlier), `execution` captures events and adds the advice in the methods themselves (i.e. not where they were called from). This automatically filters out capturing third-party library method invocations (which we are not interested in) thus making logging more efficient (unless information about which library calls are made and when is required). The `if (condition)` pointcut filters matching joinpoints using a Java condition — in our case, we filter these joinpoints using the joinpoint function name (obtained using `thisJoinPoint.getSignature().getName()`), checking whether it appears in the set of points-of-interest (`pointsOfInterest`) we want to capture. With each event satisfying this condition, we would then log the relevant information in the advice.

The level of detail one would typically log depends on the properties that will be monitored and verified and the foreseen density of such events. For the purposes of this chapter, we will be storing (i) whether the joinpoint has a `before` or an `after` modality; (ii) the method name; (iii) the target object; and (iv) the arguments. In addition, the return object is also stored in the case of an `after` joinpoint. Since we are handling all joinpoints in a uniform way, objects have to be converted to a string (from which they can later be parsed).

The event logging is implemented in the AspectJ file `Logger.aj`, which extracts events-of-interest from a whole execution and stores it in text format to a file.

Exercise 12.1

Starting with the logging code provided in the file `Logger.aj`, modify the code so that the log file also includes a timestamp for each event.

It is worth noting that as it stands, the logging program stores a trace covering *all* scenarios. Modify `Logger.aj` to create a separate log file for each scenario, by opening a new file with every call to `runScenario` and closing it upon termination.

chapter12

12.2 Replaying Events for Monitoring

Once events are captured and stored, we must build the verification component in such a manner that it consumes the events and identifies any violation that might be detected. This can be done by building a separate verification tool to evaluate the specification against the log file entries, but we can also build the verification engine using the tools that we already have. By programming an engine to *replay* the events written in the log file, we can simply use our existing runtime verification tool to monitor the replay engine. We opt for this approach so as to avoid reimplementing the verification engine from scratch.

We will start by implementing an event replayer, which parses the log file and creates dummy method calls corresponding to the original events which observed to have happened in the system. In order to avoid having to recreate the class structure of the original system, the events will not be *identical* to the original joinpoints, but rather they will use a data structure which contains all the captured information about each joinpoint. For example, an event corresponding to the joinpoint before a call to `UserAccount.withdraw(100)` would be transformed to a joinpoint data structure containing, amongst other things the type of joinpoint (before), the class (`UserAccount`), the method name (`withdraw`) and the parameters (100).

The replay engine consists of two classes: `Replayer` and `Joinpoint`. The `Replayer` class provides a constructor `Replayer(String filename)`, which reads and parses the log file, keeping track of the sequence of `Joinpoint` objects representing the recorded events. The class also includes a `replay()`

method which iterates through the events found in the log, calling the `replay()` method found in the `Joinpoint` class. This `Joinpoint replay()` method is essentially a dummy method used solely for the verification engine to observe the events.

Exercise 12.2

Review the code provided in `Replayer.java` and `Joinpoint.java`.

Extend the `Joinpoint` class to handle the timestamp field added in Exercise 12.1.

chapter12

Due to the differences between the joinpoints occurring on the system and the ones generated during the replay, the specification scripts we wrote for the runtime verification of FiTS will not work out-of-the-box with the replayer. For instance, recall the EGCL specification we had written to verify property *"The transaction system must be initialised before any user logs in"* (Property 2):

```
after BackEnd.initialise() |
  -> { Verification.fitsInitialisation(); }
before UserInfo.openSession() | !Verification.isInitialised()
  -> { Assertion.alert("P2 violated"); }
```

In order to capture the same property when monitoring the replayed events, the script would have to be modified as follows:

```
before *.replay() target (Joinpoint joinpoint)
  | joinpoint.matchFullName("*.BackEnd.initialise") &&
    joinpoint.getJoinpointKind().equals(JoinpointKind.AFTER)
  -> { Verification.fitsInitialisation(); }

before *.replay() target (Joinpoint joinpoint)
  | joinpoint.matchFullName("*.UserSession.openSession") &&
    joinpoint.getJoinpointKind().equals(JoinpointKind.BEFORE) &&
    !Verification.initialised
  -> { Assertion.alert("P2 violated"); }
```

Note that although these rules are excessively verbose, they can be generated automatically from the original rules. In fact, we can provide a translation from EGCL to offline monitoring AspectJ as an alternative version of the `toAspectJ()` method of the `EGCL` class.

Exercise 12.3

In the EGCL class, rename the `toAspectJ()` method to explicitly refer to the fact that it will produce code for online monitoring: `toAspectJOnline()`, and create a similar method `toAspectJOffline()` which generates AspectJ code to match the dummy method calls generated by the replayer instead of the ones coming straight from the system. The main difference will lie in how the event of an EGCL rule is transformed to AspectJ matching a `replay()` event with particular parameters derived from the event itself. For the sake of this exercise, keep the condition and action in the offline AspectJ as specified in the EGCL, even if they refer to variables (e.g. method arguments or return value) which would not work without further change. In order not to complicate the translation unnecessarily, you may assume that the EGCL script will have no replicated rules (`foreach` constructs).

Take the EGCL specifications you wrote in Chapter 6 for Property 2 (*"The transaction system must be initialised before any user logs in"*), and use your new code to runtime verify the property in an offline manner.

Make sure that `Logger.aj` is updated to record all events relevant to this property before attempting to offline monitor them. In practice, the changes to the logger can also be done automatically from the EGCL specification, but doing so is beyond the scope of this book.

Consider the EGCL specification of Property 8 (*"The administrator must reconcile accounts every 1000 attempted outgoing external money transfers or an aggregate total of one million dollars in attempted outgoing external transfers (attempted transfers include transfers requested which never took place due to lack of funds)"*). Keeping track of the aggregate total of external transfers requires access to the `amount` arguments of the `USER_payToExternal` method. Manually modify the generated offline monitor to access this argument from the `Joinpoint` object.

chapter12

12.3 Dealing with System State

You may have noticed that the properties we chose for offline monitoring until now depend on the method calls and their parameters. Properties may, however, refer to the system state, reading it (in the condition or action of the EGCL rules) or even writing to it (in the action part of the rules). Actions writing to the system state are impossible to handle in a purely offline setting since the verifier may have no way of accessing the system, and even if it had, the system would have long advanced. In the case of reading system state — properties which require reading information which can only be accessed from the system — offline verification is possible, but not straightforward to handle.

Consider Property 1:*"Only users based in Argentina can be gold users"*. In order to verify whether changing the type of a user to gold violates the property, we have to check the country the user is currently registered from:

```
before *.makeGoldUser() target (UserInfo userinfo)
  |  !userinfo.getCountry().equals("Argentina")
  -> { Assertion.alert("P1 violated"); }
```

Since the `getCountry()` method call is not available in the offline version, and even if it were, the country of registration of the user may have changed in the meantime[1]. In order to handle such properties, we must store all required elements of the system state with each event. There are different options available; the naïve implementation involves keeping a full system snapshot with each event, but this is obviously too expensive and not feasible in practice. In order to reduce this cost, one option is to keep track solely of the parts of the system state which are required for verification — in the case of the rule above, we would replicate the country information onto the offline system and have the system log any changes to this state so that the offline system has the latest value of the state. Yet another option is to extend the information available with the events to include the value of the parts of the system state which are required. In this case, when logging the call to `makeGoldUser()`, we would also log the value of the target user's country of residence. This last solution is ideal in terms of overheads, but can be intricate to implement automatically since it would, amongst other things, entail parsing the condition and action parts of the EGCL (which are written in Java), to deduce what data is required and where.

Exercise 12.4

The following exercises will help the reader understand better how to perform offline monitoring of a property that uses the system state such as Property 1(*"Only users based in Argentina can be gold users"*):

- Add a method `USER_changeCountry()` in the `FrontEnd` class (and a corresponding method in the `UserInfo` class) allowing a user to change their country of residence.
- Write an EGCL specification (intended for online monitoring) which runtime verifies Property 1 (*"Only users based in Argentina can be gold users"*) taking into account the new functionality. Since the

[1] As it stands, FiTS does not provide the functionality to change country of registration.

specification will be used for online monitoring, you may freely use methods such as `getCountry()` and `isGoldUser()` in the conditions.
- Extend the `Joinpoints` class to be able to parse additional key-value pairs and manually modify the AspectJ code used for logging (`Logger.aj`) to write an additional parameter storing the result of `getCountry()` whenever a user is being upgraded to a gold user and the result of `isGoldUser()` whenever the user's country is set. Then manually modify the EGCL specification from the previous exercise to process `Replayer` events but taking into account the country and gold-user parameters when available.
- Change the property such that the country of residence matters only for the first 24 hours of being given gold status i.e. gold status may only be given to users from Argentina, and their country of residence may not change for the first 24 hours. Express this property in EGCL using timers, and then manually modify the specification to be monitorable offline. What if the property enforced the country of residence to be set to Argentina within 24 hours of being awarded gold status? How would that modify the EGCL specification for online and offline monitoring?

chapter12-systemstate

12.4 Asynchronous and Offline Verification: The Fine Print

In this chapter we have explained how one can adapt the monitoring and verification techniques we have built in order to enable offline analysis of logs. As we discussed, offline monitoring is an extreme form of asynchronous runtime verification where the system and the verification proceed independently of each other, with the system writing events to a buffer, to be consumed by the verifier. In the case of offline verification, the buffer is the logfile. In this section we will discuss considerations one has to take into account when using asynchronous runtime verification.

Memory requirements: Allowing the monitor to fall behind the system requires a buffer to store excess system-generated events. Running the verification independently of the system, for example on a separate thread will still induce additional memory overheads due to the buffer. The more the verifier lags, the bigger the memory requirements. In the case of offline monitoring, the size of the logfile grows indefinitely (unless the entries are purged when consumed by the verifier) but it is made up for by having the verification take place completely independently of the system, typically on a separate machine.

Reparations: When asynchronous monitoring detects an issue, the system would typically have already progressed further — with its state changed, locked resources released, possibly even terminated. By this time, the options for reparatory actions are usually limited — the offending action cannot be stopped and the state of the system may be impossible to fix. However, it might still be possible to stop similar future actions from happening e.g. by blacklisting the user who exceeded a limit, or to perform a compensation e.g. by performing a reverse transaction (as discussed in the previous chapter).

Non-sequential access to events: Our approach to implement offline monitoring consumed events sequentially in chronological order. For some properties it may be more efficient to store the events in a database and using queries to check for violations. For instance, in the case of Property 2:*"The transaction system must be initialised before any user logs in"*, one can simply write a query to fetch calls to fetch the earliest `initialisation()` event and check whether there are any `openSession()` with an earlier timestamp, which would be more efficient than sequential consumption of events. Note that not only do we not consume the events individually, but we are checking the property in reverse chronological order (we check that there is no *earlier* event).

Initialisation of verification: Since offline runtime verification incrementally appends to the log, this may grow arbitrarily large. If verification is intermittently carried on the log, this process will become increasingly time consuming particularly since the verification process has to consume the log from the beginning every time. In order to address this issue, one would ideally have means to save and restore verification points. This can become rather intricate as the verification state would also have to be saved. Consider, for instance, Property 7:*"No user may request more than 10 new accounts in a single session"*. If one verifies a day's log, at the end of which there is still an open session in which eight accounts have been requested, the following day one would want to start from new log entries, but keeping in mind the state of verification of the property. This should ensure that on receiving three new account requests the next day, a violation should be detected.

Real-time properties: Real-time properties can be monitored in an asynchronous or offline manner by keeping timestamps associated with each event. This will, however, require the manipulation of timers in order to enable fastforwarding the simulation time forward without having to trigger actual delays. Another challenge is that of handling of upperbound properties such as Property 14:*"An administrator must reconcile accounts within 5 minutes of initialisation"*. Unless we are careful, if the system terminates before the five minutes elapse, the monitor would not be triggered. One solution in the offline setting is to have a special `end-of-execution` event to trigger any pending timer events.

12.5 Conclusions

Offline runtime verification provides a less intrusive alternative to the online counterpart by reducing the instrumentation of new code just to event logging. The drawback is that while offline monitoring still detects property violations, this will occur after the system has proceeded further to a stage where consequences of the violation may not be easily addressed.

We have seen how we can reuse much of the runtime verification machinery we built for online monitoring to perform offline runtime verification. However, one must pay particular attention consideration to certain aspects so as to ensure that the specification is correctly monitored offline.

Chapter 13
Other Advanced Topics

In this book we have focussed on exposing algorithms and implementation challenges related to runtime verification — we have looked at means of instrumentation, different verification algorithms for different specification languages and architectures for runtime verification. But there are many aspects of runtime verification that we did not touch. The intention of this chapter is to make the reader aware of a number of such topics which are deemed to be of importance.

Four topics are covered in this chapter. Firstly, we look into the issue of runtime verification efficiency concerns — how to quantify the impact of runtime verification on the performance of the underlying system and how we can minimise such an impact. The second part looks at how runtime verification can help testing, and the benefits of combining them together. Thirdly we look at techniques to make runtime verification algorithms resilient to the fact that systems may crash and be restarted, and finally, we take a look at the challenges of monitoring distributed or component-based systems and different architectures that can be used to enable this.

13.1 Efficiency Considerations in Runtime Verification

Monitoring at runtime and checking for violations comes at a cost, and since, some of the computational resources for verification are shared with the system-under-scrutiny, this is something that needs to be taken into account. Given the increase in computing power, one would assume that for the vast majority of real-world systems this is no longer a major issue. However, in reality it is not so straightforward. For instance, transaction systems rarely include computationally complex algorithms or even memory-intensive ones in the live part of the system, so one may assume that this is not really a worrisome issue for systems the likes of FiTS. However, user behaviour in such systems is rarely evenly spread over time, with high peaks typically clustered

C. Colombo, G. J. Pace, *Runtime Verification*,
https://doi.org/10.1007/978-3-031-09268-8_13

in short time intervals. For instance, a betting website typically sees a huge increase of transactions in the minutes before major sporting events, whilst e-commerce sites see such peaks just after sales are launched. The runtime overheads arising from verification negatively impact the maximum transaction volume that the system can handle any may lead to user experience degradation. In such cases, reducing overheads is crucial.

13.1.1 Measuring Overheads

The assurances offered by runtime verification are counterbalanced by the overheads induced, and in many projects this becomes a risk management exercise in choosing which properties are worth the overheads to monitor. Consider a property which says that *"at the end of a sorting routine the elements of the array match the ones that one started off with, except that they are now in sorted order"*. Here, the runtime verification costs are substantial (when compared to the system's computation), and it may thus be decided that the risks arising from a malfunction are too low to justify such costs.[1] On the other hand, a property which says that *"whenever any records tagged as health data are displayed on screen, then the data owner or another user they have authorised must be logged in"*, is cheaper to monitor, and its violation could lead to an expensive lawsuit and loss of user trust.

We have left the nature of the overheads purposefully vague until now, but this comes in many flavours: computation time, memory requirements, communication lag, power consumption, etc. In this section we will be focussing on the first two: time and space, but it is worth keeping in mind that for certain systems other overhead measures may also be important.

In order to decide whether the use of runtime verification is justified, an important step is to quantify its impact on the system's performance. In order to do so, one should simulate the system under different scenarios before deployment in order to measure the actual overheads. Deciding on the parameters of such simulations is crucial, and this is what we will discuss next.

[1] Many factors come into play here: How mature is the code? What are the direct and indirect risks arising from a violation of the property? How expensive is the computation? We kept the discussion simple, but in practice it would be important to see what the sorting routine is used for (is it ordering the treatment order of patients in an emergency room in a hospital, or the list of ingredients in a recipe?) and whether the code is a quick hack written by a new developer or part of a well-established library used by many systems over the years.

13.1.1.1 What Should One Measure?

To estimate the overheads due to runtime verification, measurements of the system's performance with and without the runtime verification component need to be taken. Before even starting to take measurements, there are various consideration one has to take into account:

- Start by assessing at which component level to measure the overheads. Is it best to take measurements of the system as a whole, at the level of a single transaction or user session, or even at the level of single method invocation. This varies from one system to another, and would have to be assessed bearing in mind the important metrics for the system-under-scrutiny. In many cases, a combination of measures would be required.
- It is not enough to decide what is to be measured. The impact of verification of a single transaction can be done with no other users logged in and no other transactions being carried at the same time, or it can be done with tens of thousands of other users logged in, performing their own transactions. The measurements may still be about that single transaction-of-interest, but the results can differ greatly. If the property requires access to a scarce resource e.g. an external database storing information about users, then the overheads will be disproportionately high in the latter scenario. It is important to identify typical scenarios e.g. average system load, peak system load and during the computation of the reconciliation process, in order to assess the impact of runtime verification under such conditions.
- When monitoring systems that involve a human or an external system in the loop, it may be more relevant to monitor the response time as perceived by the interacting parties rather than the actual overhead. If one were to measure the whole transaction including human input, the time taken for the user to interact would overshadow the overheads.

Exercise 13.1

Imagine that you were to deploy `FiTS` as a live system handling real financial transactions: (i) assess the importance of the properties we looked at in this book; and (ii) identify conditions (e.g. number of active users and sessions) under which `FiTS` should be simulated in order to assess monitoring overheads.

Program a number of such scenarios. You may consider adding delays in different parts of the system e.g. reconciliation, and between calls to the user and administrator access methods in the `FrontEnd` class to simulate human input taking time between interactions with the interface.

chapter13-overheads

13.1.2 Measuring Time

Since monitoring and verification — particularly the latter — take time to compute, it is important to understand what temporal impact runtime verification will have. At best, the system will be marginally slowed down, but at the very worst, it can result in the violation of properties such as e.g. *"Reconciliation should never take more than one hour"*.[2]

In order to take such measurements, we can either use specialised profiling tools e.g. JProfiler and Java Visual VM to measure these timing overheads, or use our own specialised timing logic. In the following exercises we will go for the latter, using a variant of the `Timer` class (as used in Chapter 10).

Exercise 13.2

`Stopwatch.java` provides a `Stopwatch` class with methods (i) `void start()` to start the stopwatch; (ii) `long getTimeElapsed()` to return the number of milliseconds since the stopwatch was started; (iii) `void reset()` to stop and reset the stopwatch; and (iv) `void fastforward(long amount)` to fastforward time by the given amount of milliseconds without waiting.

1. Review the code and add two methods `void pause()` and `void resume()` to control the stopwatch while it is running.
2. Take some of the scenarios you defined in the previous exercise and adding stopwatch control to appropriate places in the code. If you included human interaction, you can choose to pause and restart the stopwatch so as not to include human interaction delays in the timings, since these would swamp the time of the parts we are interested in. Run the simulation with and without runtime verification and measure the runtime verification temporal overheads as a percentage of the actual execution time.
3. You will have noticed that the timings taken are very low, thus leading to large measurement errors. Measure multiple executions of the same unit e.g. measure the time taken to execute multiple runs of a transaction, and divide by the number of repetitions to obtain better measurements.
4. Timings may fluctuate due to other running processes. In order to have more reliable measurements, take multiple readings by running the scenario a number of times.

chapter13-overheads

---

[2] It can even have the opposite effect, where a property such as *"Response time to notify the user of a bad password should take at least three seconds"*, is not violated due to verification overheads. Either way, this is not desirable.

13.1.3 Measuring Memory

Apart from overheads in terms of execution time, another important form of overheads are those affecting memory use. Monitoring state takes up additional memory, and we can use native Java calls to quantify this. In order to get the total memory allocated to the system being executed, we can use: the following functions `Runtime.getRuntime().totalMemory()`, while to get a measure of how much of this memory is free, we can use: `Runtime.getRuntime().freeMemory()`. However, we have to keep in mind that due to the way Java memory management works, the amount of free memory reported will depend on the garbage collector.

Measuring the memory taken by the monitors is thus not so straightforward and since we cannot force garbage collection, memory measurements may be overestimated due to monitoring state which is no longer in use but has not yet been garbage collected. In order to attenuate this issue we can *suggest* to the Java Virtual Machine (JVM) to perform garbage collection by calling `Runtime.getRuntime().gc()` before calculating how much memory is currently in use. Keep in mind, however, that the JVM may still decide to postpone garbage collection to a later, more opportune, time. Profiling tools, such as those mentioned in the previous section, can provide more precise information.

Exercise 13.3

Run one of the scenarios you built in Exercise 13.1 and measure the amount of memory used by the system with and without runtime verification.

chapter13-overheads

13.1.4 Reducing Overheads

The memory overheads due to the monitoring state being stored alongside the system at runtime is particularly pronounced in the case of parameterised properties, since monitoring memory is allocated for each object. Consider a property which says that *"a user must perform at least three incoming transfers before being given gold status"*. This can be expressed in EGCL as follows:

```
foreach target (UserInfo u)
  keep (Integer countTransfers defaultTo 0) {
  after UserInfo.depositTo(..) |
    -> { countTransfers = countTransfers + 1; }
  before UserInfo.makeGoldUser()
    |  countTransfers < 3
    -> { Assertion.alert("Property violated");
}
```

Recall that the replicated monitoring memory variable `countTransfers` becomes a hash table mapping `UserInfo` to integers using a `HashMap` object.

One first issue to note is that `HashMap` in Java uses strong references i.e. if a `UserInfo` object is no longer referenced anywhere in the system, the fact that it appears as a key in the hashmap stops it from being garbage collected. In order to address this we would have to use *weak references*[3] on the monitoring-side. Weak references do not stop the object from being garbage collected, thus ensuring that memory is freed as desired.

Secondly, through a brief analysis of the property, we can see that if the number of transfers reaches three, the property may no longer be violated, and we might as well dereference it to allow for the monitoring state to be garbage collected. If we add a way for the user to notify the runtime verification tool that the property is satisfied and can never again be violated, then it can be dereferenced. We can add a keyword `SATISFIED` to be used as a special action to allow this:

```
foreach target (UserInfo u)
  keep (Integer countTransfers defaultTo 0) {
  ...
  after UserInfo.depositTo(..)
    |  countTransfers >= 3
    -> SATISFIED
}
```

Once a property is satisfied, the monitoring code can remove the entry in the hash table for the `UserInfo` object, thus allowing the monitoring memory to be garbage collected when the system no longer references that user object. This can, however, cause problems if we observe another method on that same `UserInfo` object after stopping monitoring it. This would lead to the monitor restarting, possibly resulting in undesirable violation reports e.g. a user who made three incoming transfers will be removed from the hash table, but a request to change their status to gold after that will create another monitor instance with `countTransfers` initialised to 0; thus resulting in a violation report. In order to address this issue, in this case we would want to stop the

[3] See `java.lang.ref.WeakReference` or, more specifically for the `Hashmap` class, `java.util.WeakHashMap`.

`UserInfo` object from being monitored again. This can be achieved by keeping a set of weak references to user objects that no longer need monitoring and ensuring that the property only triggers for objects not in this set.

Exercise 13.4

1. Optimise the EGCL rule specifications to use `SATISFIED` where applicable.
2. Reengineer your EGCL tool to cater for `SATISFIED` using the approach explained above. Since actions are stored as strings it will suffice that you compare the action to the string `"SATISFIED"` to check whether this special case should trigger.
3. Rerun one of the scenarios you built in Exercise 13.1. Measure any reduction in the amount of memory used by the monitors using this optimisation.

chapter13-overheads

Needless to say, the responsibility lies with the specification writer to ensure that properties are no longer verified if they can no longer be violated. Take one of the properties we have already looked at — Property 6: *"Once greylisted, a user must perform at least three incoming transfers before being whitelisted"*:

```
foreach target (UserInfo  u)
  keep (
    Boolean greylisted defaultTo false,
    Integer countTransfers defaultTo 0
  ) {

  ...
  before UserInfo.whitelist()
    | greylisted && countTransfers < 3
    -> { Assertion.alert("P6 violated"); }
}
```

The specification writer may note that this is similar to the property we saw earlier, and after observing sufficiently many incoming transfers, the property can no longer be violated:

```
after UserInfo.depositTo(..)
  |  countTransfers >= 3
  -> SATISFACTION
```

However, with our implementation of `SATISFIED`, we excluded future monitoring of the same `UserInfo` object. But the same user may be greylisted again in the future, and the counting must start again. One way of addressing this is to have a way of stopping monitoring but only until another greylisting event occurs.

Exercise 13.5

Assess how the previous solution can be extended to have two different satisfaction actions: one which stops that object from ever being monitored for the property again (`SATISFIED-FOREVER`), and one which allows for a monitor for that object to be created if relevant future method calls to that object are observed (`SATISFIED-FOR-NOW`).

chapter13-overheads

Any experienced software engineer will probably be frowning upon seeing the solutions we have just presented. Optimisations should not come at the cost of substantial increase in complexity that can lead to bugs, even more so in verification code. However, as frequently happens in computing, what comes to our defence is abstraction.

One recurring theme of this book is how we can specify properties in increasingly abstract specification languages. EGCL was one step up from directly using aspects, symbolic automata abstracted some detail away from EGCL specifications, whilst regular expressions and LTL allowed more structure in our property specifications.

Handling satisfaction of properties written as automata can be handled by analysing which states cannot lead to a bad state. For instance, we had used the automaton shown in Figure 13.1 as a specification of Property 2: *"The transaction system must be initialised before any user logs in"*.

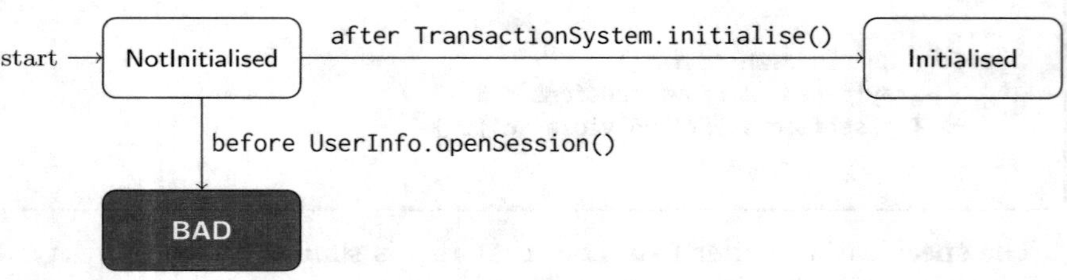

Fig. 13.1 Automaton specification of Property 2.

A quick analysis of the automaton will show that state Initialised can never lead to the bad state, and we can thus switch monitoring off once we reach it. The beauty of this approach is that it requires no intuition from the specification writer's side, but follows directly from the specification.

With the derivative-based approaches we used for regular expressions and LTL, we can use algorithmic means of deciding whether the property can ever be violated again. If we use positive specifications i.e. the formula should hold for the observed trace, we can stop monitoring once the derivative hits on a formula that can never be violated e.g. the regular expression `any*`, or the LTL formula `next true`. On the other hand, if we are writing negative specifications, i.e. the observed trace should never satisfy the formula, we stop monitoring once we hit on a derivative that can never be satisfied e.g. the regular expression `any;nothing`, or the LTL formula `always false`. Depending on the logic, such predicates may not always be efficiently computable, and they may have high algorithmic complexity or even be undecidable. In the case of undecidable predicates, we may have to use a safe approximation — if it returns true, then the property can certainly never be satisfied, but if it returns false we cannot be sure. We use satisfaction of this predicate to cut the algorithm short, so not having an exact algorithm would, at the very worst, delay the optimisation but never apply it incorrectly.

Exercise 13.6

Review the code of the `Automaton` class used to represent symbolic automata in Chapter 7 and implement a method which computes which states can never lead to a violation. Keep the analysis limited to checking whether there is a chain of transitions (ignoring conditions and actions) that leads to the bad state. If no such chain exists, the state is a safe one (i.e. it cannot lead to a violation). It is worth noting that this is an instance of an approximation — states marked as safe are certainly safe, but there may be states which are safe because of an unsatisfiable condition, but will not be marked as such e.g. a state with only one outgoing transition with condition *false*. **Hint:** It may be easier to compute the set of states that can lead to the bad state and then take the complement. To compute this set of states you can start with the set consisting just of the bad state, then repeatedly add all states that have an outgoing transition to a state in this set. Stop when no states remain to be added.

In the case of negative properties specified using regular expressions we need a predicate to decide whether a particular regular expression is no longer satisfiable — we already defined such a predicate *cannotMatch*() (see Section 8.4). Modify the verification code handling the monitoring of parameterised negative regular expression properties to switch off monitoring of a formula once this predicate is satisfied.

chapter13-overheads

13.2 Persistent Runtime Monitors

One aspect of runtime verification which we have not discussed is the persistence of runtime monitors. As with any other parts of the system state, if such monitors are exclusively kept in memory, they are lost the moment the system stops running, and resuming the system would restart the monitors from scratch, thus possibly diverging from their originally intended behaviour. In many real-life scenarios this is not desirable, and monitors must be persisted for several reasons: (i) the system has a persistent state and as such the properties span multiple system runs (e.g. FiTS using a database to store user information, and set up such that it can be shut down and restarted at will); (ii) the system (including the monitor) may crash and need to be restarted; (iii) the monitoring component might crash independently of the system (if this is kept separate from the system). Once the system and monitor is back up, the monitor should be able to continue from where it left off, which is only possible if the monitoring state is also stored in secondary memory. This may also be required if the size of the monitor grows too large to keep completely in memory.

This section takes the reader through the basics of monitoring state persistence, starting with hand-coded instructions within the rules[4], to more sophisticated approaches which involve a fully fledged database and caching mechanisms.

A number of FiTS properties do not require the saving of any state, e.g. Property 1:*"Only users based in Argentina can be gold users"*, and Property 3:*"No account may end up with a negative balance after being accessed"*, only check the system state at a point in time to decide compliance or otherwise. However, the situation is different for other properties which require the use of a monitoring state, e.g. Property 2:*"The transaction system must be initialised before any user logs in"*, and Property 6:*"Once greylisted, a user must perform at least three incoming transfers before being whitelisted"*.

Focusing on the latter example, recall that we used two variables to keep track of the monitoring situation at any particular point in the system execution: whether the user has been greylisted and the number of transfers carried out since the greylisting. To save the monitoring state, we must store the values of these variables in secondary memory to be able to recover them in case we need to resume monitoring at a later stage.

We can resolve this through a manual saving and restoring of the monitoring state variables using methods `MonitoringState.saveValue()` and `MonitoringState.restoreValue()` in the action part of EGCL rules, which is saved to and loaded from secondary storage as required. Sticking to the greylisting example, if we want to ensure the monitor maintains its state between separate runs, we can trigger on the transaction system's `initialise()`

---

[4] As noted earlier, we will refer to EGCL rules in this section for simplicity but the extension can be also applied to other specification notations.

and shutdown() methods to ensure the monitoring state is loaded and saved as required:

```
before TransactionSystem.initialise() |
  -> {
    MonitoringState ms = new MonitoringState("filename");
    Verification.isGreylisted =
      (Boolean)(ms.restoreValue("isGreylisted", false));
    Verification.countTransfers =
      (Integer)(ms.restoreValue("countTransfers", 0));
  }

after TransactionSystem.shutdown() |
  -> {
    MonitoringState ms = new MonitoringStore();
    ms.saveValue(
      "isGreylisted",
      Verification.isGreylisted
    );
    ms.saveValue(
      "countTransfers",
      Verification.countTransfers
    );
    ms.saveToFile("filename");
  }
```

The second parameter of the restoreValue method takes a default value to use in case no value has yet been saved. By storing the data in secondary storage e.g. by serialising or converting the monitoring state to JSON format and storing the data in a file.[5]

Exercise 13.7

Review the code provided to see how MonitoringState works internally, then use the methods to restore (MonitoringState.restoreValue()) and save (MonitoringState.saveValue()) the memory state to keep track of the state of Property 8:*"The administrator must reconcile accounts every 1000 attempted outgoing external money transfers or an aggregate total of one million dollars in attempted outgoing external transfers (attempted transfers include transfers requested which never took place due to lack of funds)".*

chapter13-persistence

Since manually saving the individual monitoring variables can be cumbersome and error-prone, one can organise the monitoring state into a separate

[5] The implementation provided uses serialisation.

object and save the object as a whole. We can even go a step further and have the monitoring state explicitly declared in the EGCL script in order to automatically synthesise global save and restore methods. This works well when we save and restore the state at select points as we have done here.

However, one weakness of the way we are saving and restoring state is that the monitor is only saved when the system terminates normally. If the monitor or the whole system crashes, the monitoring state is not saved and thus cannot be resumed upon being restarted.

One way around this is to save the monitoring state at the end of every single rule action. Whether one does this manually through an explicit call to the method to save the whole state, or whether it is integrated automatically when instrumenting the monitor, one has to be careful: (i) saving the whole state at the end of every rule is overkill and one should, instead save only the part of the state that may have changed; (ii) saving the information to secondary memory with every event is expensive and should ideally include a caching mechanism to reduce overheads.

Without such a caching mechanism, the approach does not scale up to industrial systems where high traffic volumes require sophisticated optimisations of persistent memory management. In such cases, the monitor should reflect the system's persistence mechanism: any update to the system state in persistent storage should be reflected in the persistent state of the monitor. Similarly, the caching mechanism used by the system for efficiency should ideally also be reflected in the monitoring infrastructure. Given this correspondence between the monitor and the system, it would be ideal that if monitoring is done online, it is included in the design of the overall architecture right from the start.

13.3 Testing and Runtime Verification

Testing shares much with runtime verification: both techniques involve the actual execution of the system-under-scrutiny, and both techniques check for the correctness or otherwise of a particular trace. And yet, they are also miles apart. Firstly, testing is performed during the development of the system, whereas runtime verification is typically carried out on the live system post-deployment, and secondly, whereas runtime verification checks a real user-driven runtime trace, testing typically checks a synthetic trace created by the test engineers (or using a trace generation algorithm).

If we were to characterise the two techniques, we could say that runtime verification involves *event elicitation* and *behaviour checking*, whereas testing involves *test case creation* and *behaviour checking*. In this section we explore how runtime verification techniques can be useful for testing.

One of the more common approaches to testing is *unit testing* — applying testing to small units of the system isolated from the rest, possibly even producing substitute code e.g. mock objects, to simulate the rest in order to ensure such isolation. Typically, testing consists of a number of such unit tests, where each test is a list of method calls invoked on a newly set up system, followed by an assertion about the resulting state of the system.[6] If an assertion is not satisfied, the test is considered to have failed.

In order to get the most out of this section, the reader ideally already has a degree of familiarity with JUnit, a testing framework for Java. If you do not, it would be helpful to install JUnit and follow a tutorial before reading further.

A JUnit test class essentially is a normal Java class, but with particular test annotations. Consider the test class below to test that after the administrator initialises FiTS, creates a user and enables them, then the user is indeed enabled:

```
public class FiTSTest {
  static TransactionSystem ts = new TransactionSystem();

  @BeforeEach
  void setUp() throws Exception {
    ts = new TransactionSystem();
    Verification.setupVerification();
  }

  @Test
  void testUserEnabled() {
    FrontEnd fe = ts.getFrontEnd();
    fe.ADMIN_initialise();
    int uid = fe.ADMIN_createUser("Fred", "France");
    fe.ADMIN_enableUser(uid);

    assertTrue(fe.getUserInfo(uid).isEnabled());
  }
}
```

The `@BeforeEach` annotated method in the class provides the logic to reset the object before each test, while the `@Test` method carries out the execution of the test trace ending with a JUnit assertion.

Exercise 13.8

Create another test which sets the created user's status to silver, and asserts that the status has indeed been updated.

chapter13-testing

Unit testing is ideal for testing the behaviour of a method, but is less effective when the testing scope grows wider. Expressing a system-wide property

[6] In testing, such an assertion is typically referred to as an *oracle*.

as a basic assertion is not always straightforward to do since, as we saw in Chapter 4 when we were expressing runtime verification assertions manually into our code, for any non-trivial property we end up adding additional information and logic, resulting in increased complexity in the testing code.

Using runtime verification, we can support testing by providing more expressive and abstract specification languages which are automatically compiled into monitors. Model-based testing can be used to generate paths of interest through the system state space, while using runtime monitors to express oracles for sophisticated properties expressed using an appropriate formalism.

For instance, we can create tests to check for possible violations of properties without specifying the oracle assertions. For instance, a test case for Property 1:*"Only users based in Argentina can be gold users"*, is the following:

```
public class FiTSTest {
  static TransactionSystem ts = new TransactionSystem();

  @BeforeEach
  void setUp() throws Exception {
  //equivalent to Scenarios.resetScenarios();
    ts = new TransactionSystem();
    Verification.setupVerification();
  }

  @Test
  void enableUser() {
  //equivalent to Scenarios.runScenario(1);
    ts.getFrontEnd().ADMIN_initialise();
    int uid = ts.getBackEnd().ADMIN_createUser("Fred", "France");
    ts.getFrontEnd().ADMIN_enableUser(uid);
    ts.getFrontEnd().ADMIN_makeGoldUser(uid);
  }
}
```

We can then run the test but with the system monitoring the property specified as EGCL rules, using our runtime verification tool:

```
after UserInfo.makeGoldUser() target (UserInfo u) |
  -> { assertTrue(u.getCountry().equals("Argentina"), "P1 violated"); }
```

By using the JUnit `assertTrue()` method rather than `Assertion.alert()` we ensure that the JUnit test fails. In this manner, we are separating the writing of the test drivers from the oracles (which can become quite complex), and use a runtime verification tool to run the oracle. The other advantage of this approach is that oracles deemed to be important can easily be kept post deployment, simply by changing `assertTrue()` to appropriate violation handling code.

Exercise 13.9

Run the test shown above with the runtime-verification-programmed oracle.

Write a test class for testing adherence to Property 5: *"Once a user is disabled, he or she may not withdraw from an account until the administrator enables them again"*. Run the test scenarios with the EGCL runtime verification tool, such that a JUnit violation will be reported during testing if the system fails in the test scenario being run.

chapter13-testing

13.4 Monitoring Architectures

In this book, we have assumed that the system-under-scrutiny is a single monolithic system. This gave us two options: integrating the runtime verification part as part of the system (as we did in most of the book), or separating it as an independent component (as we did in Chapter 12). When a system is composed of different independent components, we have more options and we face new challenges. Whether the system is made up of different components, whether it is split into parts built using different technologies, or in any other way, we have to decide on issues such as how to write specifications spanning across different parts of the system, and where runtime verification code should reside.

13.4.1 Distributed FiTS

In order to motivate better this section, we will present DistFiTS, a simple distributed version of FiTS consisting of two largely independent instances of FiTS which can communicate together. DistFiTS can be seen to represent two online branches of the same financial institution, sharing users (through their user id, but not necessarily synchronising their status, type and mode), but with independent accounts and sessions i.e. a money account and a session may reside on one or the other of the instances, but not on both.

The properties we have seen can be adopted for the single instances of FiTS. However, we may now have properties spanning across the two instances. For instance, we may have an overarching property across the instances to ensure that we do not violate a regulatory requirement that across all its branches, the system should not handle more than a certain amount of money:

Distributed property 1: *"The balance of accounts held by the two instances of FiTS together should not exceed $10,000,000".*

We can also have properties which temporarily span across the different instances, such as a property asserting a weak form of blacklisting migration:

Distributed property 2: *"Once a user is blacklisted on one of the instances of FiTS, they must also be blacklisted on the other instance of FiTS by the end of its next reconciliation".*

Properties may constrain behaviour across instances of FiTS:

Distributed property 3: *"The first FiTS system must be initialised before the second".*[7]

In order to experiment with this setting, you will find a version of DistFiTS in chapter13-distributed consisting of two instances of `TransactionSystem` and a means for them to communicate together:

- `DistTransactionSystem.java` models the distributed transaction system. It consists of two instances of the `TransactionSystem` class, both aware of each other. You should not think of this class as a real component of DistFiTS, but rather as a means for us to simulate the two independent instances of FiTS running together.
- `TransactionSystem.java` provides a number of new methods: (i) the method `void register(TransactionSystem ts)` registers the existence of the other instance of the `TransactionSystem` with this one; (ii) the method `TransactionSystem getOtherTransactionSystem()` returns the other instance of `TransactionSystem` registered (returning null if it has not been registered); (iii) `void message(String m)` is the method which will be called by other components (in particular the other instance of the `TransactionSystem`) to send us a message. Initially we will ignore messages, but will add code to process them later on.
- The classes in `TransactionSystem.java`, `BackEnd.java`, `UserInfo.java`, `BankAccount.java` and `UserSession.java` are augmented with a new method used to send a message to the other instance of the transaction system (by calling the other system's `message` method): `void sendToOtherInstance(String msg)`.

Needless to say, modelling DistFiTS in this manner has its limitations. Essentially we are modelling a non-monolithic system as a monolithic one running on the same address space. But keep in mind that this is purely for modelling purposes, and the `DistTransactionSystem` object should be seen as a modelling construct, and not an actual object in the implementation of DistFiTS. Also, the two instances of FiTS are conveniently identical and built using the same technology, which is not always the case in real-world systems. Finally, the communication channel is implemented in a simple synchronous

[7] We assume that the two instances are distinguishable allowing us to say which one must be initialised first.

manner — when one instance of FiTS sends a message to the other instance (using the latter's `message()` method), it will immediately be received and processed.

Indeed, we can use our runtime verification tools to instrument properties on the `DistTransactionSystem` object, but that would be cheating, since we would not have had that luxury had the two FiTS systems been really distributed. We will thus refrain from doing so, and will be looking at ways of instrumenting the two instances separately. Similarly, we will refrain from having the FiTS implementations from calling each other directly except through the message sending mechanism thus ensuring that the distributed model is adhered to.

Exercise 13.10

Review the code provided, and make sure that you understand how DistFiTS works. Implement the following extensions:

1. The implementation of the instances of FiTS has to be extended in order to be able to deal with the distributed properties. We will now add logic attempting to address this:
   - Extend the implementation of FiTS such that (i) it keeps a balance of all its accounts, and (ii) it memorises the total balance of the other instance of FiTS. When a transfer violates the \$10,000,000 limit, simply write a warning to the standard error stream.
     Upon every successful change of an account balance, have the transaction system instance where that change occurred send a message to the other instance notifying it of the change. Since only string messages are allowed, encode the message appropriately e.g. `"T-2.42"` would represent a transfer that reduced the total balance by \$2.42, whereas `"T+1.23"` would represent an increase of \$1.23. For simplicity, you may assume that transfers are never between accounts one of which is on one instance of `TransactionSystem` and the other of which is on the other instance.
     Add message handling to update the stored balance of the other FiTS instance whenever a message is received.
   - Add a message to be sent whenever a user is black-, grey- or whitelisted on a `TransactionSystem` instance. Implement the message handling method so that FiTS keeps track of such received messages and blacklists a user upon their first transfer attempt if they were blacklisted (and not subsequently white- or greylisted) in the other FiTS instance. Keep in mind that you may assume that user ids are shared between the two instances.

2. For each of the two distributed properties, define a scenario violating that property. In the case of the first property, note that we only reported that the maximum balance was exceeded, whereas with the second property, the suggested implementation in the previous question does not satisfy the property. Ensure that your scenarios exhibit such violations.
3. The messaging mechanism implemented in the DistFiTS model is unrealistically synchronous. Implement an alternative version of DistFiTS changing messaging to simulate asynchrony as follows:
   - The `message` method will now simply store messages in a buffer (a queue).
   - Add a method `void consume()` which takes the first message in the buffer and consumes it.

   In the scenarios of this version of DistFiTS, you will have to add additional calls to have the transaction systems consume their messages. Note that, in this manner we are simulating asynchronous communication, where a message sent may not arrive instantaneously.

chapter13-distributed

13.4.2 Specifying Distributed Properties

You will have noticed that since our specification languages refer to events based on objects and their method names, for properties about distributed system, we will need means of distinguishing between events coming from different components. Similarly, if our specification language handles conditions and actions, we would need to know where they are to be evaluated. One solution to this is to tag events, conditions and actions with the location where they will be triggered. For instance, recall Distributed property 3 which says that the first FiTS system must be initialised before the second, and which can be expressed as an EGCL property as follows:

```
Boolean fitsInitialised@monitor = false;

(after FrontEnd.ADMIN_initialise())@ts1 |
  -> { ts1Initialised = true; }@monitor

(before FrontEnd.ADMIN_initialise())@ts2
  |  (!ts1Initialised)@monitor
  -> { Assertion.alert("Property violated"); }@monitor
```

Note that monitoring state variable declarations, events, conditions and actions are all tagged with a label stating where the variables are to be stored, where the event is expected from, and where the condition and action code is to be executed. In this case, we would have to declare elsewhere the two systems named `ts1` and `ts2`. The parts of the specification that are tagged with the `monitor` location are not committed to be stored and executed anywhere in particular, and it would be up to the instrumentation and verification algorithms to decide where the monitor will be placed. We will discuss this in more detail later on.

In practice, things can get more complicated than this. For instance, the violation rule might also have some code to execute at the `fits2` location to make up for the violation, and would require a compound action with different parts executed in different places. Similarly, conditions may depend on code to be executed on multiple locations and may have subexpressions tagged with different locations. A more sophisticated violation rule which permits the second instance of FiTS to be initialised before the first as long as it has no user accounts can be expressed as follows:

```
(before FrontEnd.ADMIN_initialise())@ts2
  |  (!ts1Initialised || (ts.getNumberOfAccounts() == 0)@ts2)@monitor
  -> { Verification.alert("Property violated"); }@monitor
     { FrontEnd.ADMIN_reportUnexpectedInitialisation(); }@ts2
```

As is evident from the above example, such location-aware properties rapidly increase in complexity. For instance, in the condition we had to decide whether the disjunction is to be executed on the monitor or on the second instance of FiTS. Such micro-management of property checking is not ideal.

Another source of complexity not dealt with above, is that the example we used just had two fixed locations (and one monitoring location). In practice, we may have dynamic locations, which are discovered at runtime. For instance, a peer-to-peer transaction system might allow for new nodes to register on the network and provide certain functionality. Such dynamic locations continue to increase the complexity.

In this book, we will stop at identifying these challenges without exploring solutions, but these are issues that have to be kept in mind and addressed if one is to implement monitoring of distributed systems.

Exercise 13.11

Express the two distributed properties using location-aware EGCL rules.

chapter13-distributed

13.4.3 Distributing Runtime Verification

We can now turn to considerations concerning the runtime verification of such distributed systems. Clearly, the code to capture the events must be localised to the individual components, but there are options as to where these events are consumed and where verification is performed. This is particularly important in distributed systems, where communication overheads swamp computational ones, and it is important to be aware of any additional communication due to monitoring.

When one looks at distributed systems there are, broadly speaking, two approaches as to how the different components work together. On one hand, there is *orchestration* in which (much like an orchestra) there is a central component (the conductor) leading the other components and ensuring that they are working in sync, typically having communication between components happen through the centralised orchestrator. On the other hand, one can go for a *choreographed* approach, in which each component (much like a dancer in a choreographed dance) acts independently of the others, communicating with others directly as the need may arise. The former has the advantage of being easier to coordinate, but raises concerns of having a single point-of-failure and an increased communication load. The latter is typically more complex to design correctly, but reduces communication and decentralises points-of-failure.

In the context of runtime verification of distributed systems, the two approaches are possible.

- In an **orchestrated monitoring approach**, a component, in our case a new independent monitoring component (or possibly one of the FiTS systems), will be designated as the orchestrator, and will receive all events and data of interest from all the components. All the verification logic is performed on the orchestrator, and any actions to be carried out on the other components are also triggered centrally.
  If we want online monitoring, the individual components would then have to wait for a response coming from the orchestrator before going ahead with further computation, thus being able to carry runtime reparations if a violation is discovered.
  Recall the third distributed FiTS property: *"The first FiTS system must be initialised before the second"*. In an orchestrated approach, both FiTS instances would send a message to the orchestrator whenever initialisation takes place (since these are the events of interest). The orchestrator then performs all the computation relevant to verification — keeping track of whether the first FiTS has initialised, and sending back a message to trigger reparatory action if the second FiTS was initialised before. If we look at the way we expressed this property as a location-aware EGCL property, all monitor-side data and code is stored on the orchestrator and given any

events, conditions or actions tagged with other locations, these would be executed through communication with the orchestrator.

- In contrast, a **choreographed monitoring approach** keeps verification local and communicates with the relevant component when verification has to continue elsewhere.
  With the initialisation property, you may have realised that the monitoring on the first instance of FiTS needs only to wait for initialisation to take place and then let the other transaction system instance know and stop any further monitoring. The second instance similarly has to monitor for an initialisation event, and trigger reparation if one is observed. If, however, before it is initialised it receives a message notifying it that the first transaction system initialisation took place, it can stop monitoring altogether, thus not triggering violations.
  Given a location-aware EGCL property to monitor in a choreographed manner, one has to decide where the monitoring state variables should go (possibly passing their value as monitoring proceeds), which monitors go on which system and where to perform computation when events are observed. In this case, the second FiTS instance would manage the monitoring state, keeping track whether or not the first FiTS instance has been initialised (since it is the component that needs this information), and updating it if it receives a message from the first component saying that it was initialised. The monitoring state is checked upon observing an initialisation of the second FiTS instance to decide whether or not to trigger a reparation.
  Note that this transformation of the property into different components communicating directly between themselves is much more complex than what has to take place in the case of orchestration.

As can be seen from this example, the orchestrated approach is much easier to instrument, but requires more communication and is more easily exploited.

A challenge in distributed systems that we have not yet discussed in the context of runtime verification is that of asynchronous communication. Take the orchestrated approach, with the instances of FiTS initialising almost at the same time, but still in the right order. Although the order was as desired, the messages being sent to the orchestrator by the two instances of FiTS may be received in the wrong order e.g. due to network latency. One cannot assume that events happen in the order they are received. The same problem can arise in a choreographed approach if the second instance sees an initialisation event before receiving the message that the first system was initialised due to a delay in the communication channel. This is a well known problem in distributed systems and although there are partial solutions, it is not always possible to resolve the issue no matter how the instrumentation is organised.

Exercise 13.12

Starting with the location-aware EGCL properties you wrote in the previous exercise, manually implement them using both an orchestrated and a choreographed runtime verification approach using EGCL. For the orchestrated approach create a new `Orchestrator` component in the `DistFiTS` class. The `Orchestrator` should store the monitoring state variables, have a method to allow the registration of two `FiTS` instances which will be verified `void register(TransactionSystem ts1, TransactionSystem ts2)`, and a `void message(String m)` method to enable the `FiTS` components to send information to this component. The registration method in the `TransactionSystem` class should be extended so that it also receives the orchestrator object: `void register(TransactionSystem ts, Orchestrator o)`.

The aim of this section was to show the challenges of applying runtime verification to component-based systems where properties may span across the different parts of the system. Although the running example was that of a distributed system, the principles can be applied more generally. For instance, in a heterogeneous system that is composed of a number of subsystems each of which may be implemented using different programming languages, or deployed on different technologies, the approaches we discussed are still applicable. The runtime verification tool would obviously have to be general enough to be able to instrument code in the different types of subsystems encountered and also be able to handle communication between them, but otherwise the choice of monitoring architecture, the property specification approach, etc. remain unchanged.

13.5 Conclusions

In this section, we have touched upon a number of more advanced topics in runtime verification. Although we did not go into as much detail as we did in the rest of the book, the intention was that of exposing the reader to these issues.

Chapter 14
Conclusions

Despite all efforts to eradicate bugs from systems prior to deployment, wrong system behaviour remains all too common an occurrence. In this book we have presented the ideas behind runtime verification and incrementally constructed tools to allow for the verification of runtime behaviour. Runtime verification can be seen as a fallback insurance, to ensure that any bugs not discovered pre-deployment can be discovered during the execution of the live system, allowing for reparatory action to be taken.

As with all validation and verification techniques, the effectiveness of runtime verification depends on the completeness and correctness of the specifications and properties used to verify against. Indeed, the properties written for verification may not include important wrong behaviour or may even misclassify certain behaviours — bugs in specifications can be as detrimental as the ones in the implementation itself. However, specifications typically stop at describing the ideal state of affairs (e.g. *at the end of the sorting method, the elements of the list should be in ascending order*), and do not delve into how such a state of affairs can be achieved (e.g. *use mergesort*). This not only ensures that verification can be independent of design decisions, but also results in smaller artifacts (i.e. the specification is smaller than the program) and is expressed at a higher level of abstraction. This is why the right choice of specification language is essential.

14.1 Runtime Verification Concerns

The ease of use of runtime verification makes it an attractive technique to be adopted, but there are a number of concerns that arise with its use in the industry. The major concern is one of overheads induced on the main system's performance due to monitoring and verification algorithms. Depending on the properties being monitored, these overheads may become prohibitively large, making the use of runtime verification unfeasible. Consider, as an example

C. Colombo, G. J. Pace, *Runtime Verification*,
https://doi.org/10.1007/978-3-031-09268-8_14

of an unreasonable use of runtime verification, a property which says that *"After executing the sorting method, for any indices i and j such that $i \leq j$, the array elements at those indices are also similarly ordered: $a[i] \leq a[j]$"*. Such a naïve check would induce overheads taking time quadratic in the size of the array $O(n^2)$,[1] outweighing the complexity of the sorting routine itself $O(n \log n)$. Even a more efficient implementation of this property which checks only adjacent elements $a[i] \leq a[i+1]$ requires computation of time complexity $O(n)$ which unnecessarily burdens the sorting method. To ensure that monitoring and verification is feasible, one can limit oneself to certain specification languages or parts thereof, which guarantee low overheads. However, this may not be an option, in which case a reasonable choice may be to relegate certain checks to be performed offline, or at least asynchronously with the system, thus ensuring that the system proceeds independently of the monitors.

The choice of specification language for use in the industry is crucial, requiring a balance of expressivity, ease of use, efficiency of monitoring, etc. Narrowing down on ease of use, if specifications are to be written by non-technical persons, an intuitive domain specific language (DSL) might be preferred over readily available notations. The advantage of a DSL is that it can be tailored according to the specific needs of the context, natively supporting concepts depending on the domain. For example in the case of a financial transaction system, runtime monitoring could be used to gather statistics about the system execution. In such a situation, one could require the language to support statistical concepts such as running averages, standard deviation, etc.

Overheads in time and space are not the only way in which runtime verification can interfere with the system under scrutiny. Monitoring can interfere with timing of the original system, possibly resulting in new race conditions or affect the correctness of the system. For instance, a system which ensures that a transaction never takes more than a certain amount of time to execute, may violate the timing constraint only when the monitor for the very same property is introduced. Other undesirable interactions may also include that the monitors might expose private data, thus raising privacy concerns.

Yet another concern when introducing runtime verification is a software engineering one, arising due to the fact that we are adding new tools and artifacts as part of the development process. This may raise concerns that the resulting processes are more fragile in that this adds another potential source of system failure. It is difficult to argue for the adoption of external tools, particularly academic proof-of-concept tools which may not have been developed according to accepted software engineering standards and the established practice of the development team. Our approach in this book enables the reader to build their own internal runtime verification tools specialised for the particular needs of the system-under-scrutiny in question.

[1] Here we are assuming that the check is only performed after sorting the whole array, but in practice, a recursive sorting algorithm such as quicksort or mergesort would also trigger such checks after sorting subarrays, making the cost even higher.

14.2 Adopting Runtime Verification in the Industry

Despite these concerns, runtime verification is a very industry-friendly approach in that adopting it does not require extensive additional expertise or resources. One of the challenges is that of exposing the value of runtime verification, the benefits of the approach and how it complements existing setups. Typical questions and doubts we have repeatedly encountered when introducing these techniques in industry were:

- *"We already have checks for those properties in the code. Why duplicate the effort?"* At first sight, the development of code in such a manner that it appears to adhere to the specification and adding the occasional assertion in the code to check for consistency is frequently conflated with the adoption of runtime verification as part of the development process. The dissociation of the specification and the system code is a strength of runtime verification which is easily overlooked at first. Furthermore, a systematic approach to integrating monitors in a system ensures that properties and specifications can be extended or modified more effectively.
- *"We already test for those properties."* Having tested for a property does not necessarily mean that the system will not violate it under different (untested) circumstances. Test coverage ensures that the different parts of the code are explored by the tests, but does not take into consideration all possible orderings of the events. Runtime verification addresses this by checking for correctness with the trace seen at runtime, and thus complements testing by extending checks beyond the deployment horizon.
- *"Properties? What properties?"* Frequently, developers, testers and quality assurance engineers think of systems as input-output machines whose correctness can be checked for a table of possible input and output behaviours, ensuring that the system is still functional and has produced the expected output during each test run. Generalising the notion of correctness to a symbolic description which explains the expected behaviour for any trace is not common practice. There is, however, great value in identifying general properties which should hold of any system execution path, providing an abstract notion of what the system should be doing as opposed to focussing solely on how it can be achieved.
- *"Why don't you talk to the quality assurance department/developers/system architects? They should be doing this."* Where does runtime verification fit in the software engineering process, and whose responsibility is it? All too frequently we found that many teams saw it as the responsibility of some other department — anyone but themselves. Indeed, one challenge is that verification is a cross-cutting concern and is related to what the system architects, the developers and the quality assurance engineers do. Seen in this way, it is easy to understand why all the teams see the need to involve other teams. The question of how runtime verification fits in the engineering process has unfortunately not been sufficiently studied

in the literature. From experience, it finds its most natural home within the quality assurance team but with the involvement of other teams, for instance the development team to address system invariants at a code level (as opposed to just at a business logic level) and the system architects to handle properties concerning the integration of components.

If one is seeking to adopt runtime verification on substantial projects, it is recommended to start small and go bigger later — identifying small parts of the system whose correctness may be of a more critical nature, or which may be more amenable to monitoring. Furthermore, in many cases, offline monitoring is a good way to start, since this will minimise system intrusion and allows not only for verification but also for the business teams to pose what-if questions e.g. *"What if we tweaked this parameter — would some other fraudulent cases have been caught?"* Finally, including business-level specifications typically helps exposing the benefits of the approach to management.

By now the reader should be sufficiently aware of the opportunities brought about by runtime verification to be able to assess where and how these techniques may help in their projects. All those involved in software development are aware of the risks due to software failure, and runtime verification is one more tool to mitigate some of these risks.

Appendix A
Further Reading

In this appendix, we will be giving pointers to further reading that the interested reader may want to follow. The intention of the book is that of exposing runtime verification in a hands-on and practical manner. However, the field of runtime verification is a mature one and there is substantial work that would be of interest to the reader who wants to delve deeper, whether from a theoretical or practical perspective, into the different aspects covered (and others not covered) in this book. For each chapter, we give a list of relevant articles which may be of interest to the reader.

Chapter 1

Chapter 1 focussed on the need for verification — how common failure in computer systems is, and how such failure can have a serious impact on individuals or society as a whole, and thus the need for dependability through verification or otherwise.

One can find various books and articles on computer systems failure, some aimed at emphasising the difficulty of having dependable software e.g. [69, 58], to ones which document instances of such failure e.g. [57].

Chapter 2

In the second chapter, we introduced runtime verification focussing on the practical perspective. If you are after a good introduction to runtime verification including a more theoretical take on the topic, [20, 53] are a good place to start. More recent introductions to runtime verification can be found in [35, 11].

C. Colombo, G. J. Pace, *Runtime Verification*,
https://doi.org/10.1007/978-3-031-09268-8

There are also papers presenting a runtime verification taxonomy [29, 36], both of which present a more structured view of the area and the tools which have emerged over the years.

Chapter 3

This chapter introduces a financial transaction system, FiTS, as a case study providing running examples throughout the book. Our experience with financial transaction systems [23] has shown that they provide an interesting point of entry for runtime verification into industry where the stakes are not as high as in a critical system such as a nuclear power plant, but the repercussions of something going wrong undetected still serious enough to warrant effort in monitoring.

The choice of FiTS as a running example and the kind of properties chosen set the scene for the rest of the book. The runtime verification literature includes a rich variety of domains, including: low-level properties about the correct usage of Java libraries (e.g. correct usage of iterators) [14, 49], time-sensitive systems [56], embedded systems [16] and decentralised and distributed systems [41].

Chapter 4

As a basic introduction to runtime verification, this chapter explains how runtime monitors have their roots in assertions originating from the work on design-by-contract [55].

For a historical perspective on the use of assertions in programming languages, [19] gives a good overview. A personal view on assertions from the eyes of C.A.R. Hoare can be found in [47], which is particularly interesting given his role in developing the pre- and postcondition approach to program verification, better known as Floyd-Hoare logic [46, 39].

Chapter 5

Aspect-oriented programming [51] has been proposed as a way of programming cross-cutting concerns in a modular fashion. The earliest work proposing the use of AOP for runtime verification goes back to 2002 [68] but several other works e.g. [26, 14, 49] followed, mostly using AspectJ [50].

Chapter 6

Rule-based reasoning — proposed in this chapter as a specification language — has been originally proposed by Dijkstra [31] as a structured tool for programming language semantics. Since the early days of runtime verification, rules have been proposed as a means of expressing properties [8] with Eagle [43] and RuleR [9] being two such central tools.

More recently, rule-based specifications have also been used within polyLarva [21] and Valour [5].

Chapter 7

In Chapter 7, we presented finite state automata as a diagrammatic way of expressing runtime verification properties with implicit state. Finite state automata (also referred to as finite state machines) have been used as a specification notation in the JavaMOP framework [49]. Extended versions of automata have also been proposed in other work [26, 61].

Chapter 8

The argument for structure goes back a long time, perhaps most famously when Edsger W. Dijkstra argued for the harmful nature of the GOTO statement [30]. This chapter proposes regular expression as a structured approach as opposed to finite state automata.

To monitor regular expressions, we have used derivatives — originally proposed by Brzozowski in 1964 [15] — as a way of computing the resultant expression to be satisfied given the past observation. The idea is closely related to that of term rewriting, where a set of formulae are updated based on newly available truths, and has been extensively explored [62] in the context of runtime verification.

One can read more about the use of regular expressions for runtime verification specifications in [64, 14].

Chapter 9

LTL was originally proposed by Pnueli in 1977 [60] for program verification. Since then it gained popularity for verification and it was natural for the pioneers of runtime verification to import it with them into the new field [42]. However, as a consequence of LTL semantics having originally been defined over infinite traces for verification, not all of it is monitorable, and several varying semantics emerged for use with monitoring [12]. For instance, based on a three-valued logic semantics of LTL[1] Büchi automata were used to synthesise deterministic finite state automata for runtime verification from LTL formulae [13]. Given LTL's importance in runtime verification, the concept of monitorability has been well studied [34, 1], both with respect to LTL and more generally.

Chapter 10

Real-life systems, particularly those dealing with human users, adhere to time limits to remain usable. Over the years, there have been several proposed specification languages [18, 3, 4] to describe real-time properties. This chapter provides the basic mechanisms for monitoring of such concerns, distinguishing between lower and upperbound properties. There are a number of runtime verification tools which support timed properties (e.g. Larva [26], TeSSLa [52], and Striver [44]). Toward the end of the chapter, we highlighted a number of challenges involved in checking for such properties, particularly since monitoring and verification themselves take time. This change in timing may cause bugs and/or solve others, and care (and possibly analysis [27]) must be taken to prevent this from happening.

Chapter 11

Although flagging violations is a crucial aspect of runtime verification, in several instances, one would also want to trigger some form of reaction to the observed behaviour (see [37] for an overview).

In ideal situations, the system is not even allowed to violate the properties under consideration. Known as runtime enforcement [59], this area of study deals with blocking events, or triggering others to steer the system away from unwanted behaviour. When this is not possible, i.e. the earliest detection time is later than the occurrence of the violation, other options could be considered involving some form of rollback of the system state [25]. More generally, runtime-monitor-triggered reactions can be used for modularly adding functional elements to a system. This is better known as monitor-oriented programming (MOP) with a substantial body of work [54] behind it.

---

[1] A finite trace may suffice to show that an LTL property to be *true*, or to be *false*. However, it may not provide sufficient evidence of either, thus resulting in a third value *don't know*.

Chapter 12

For several reasons, but most typically to keep monitoring intrusion to a minimum, offline runtime verification might be preferable, or indeed the only viable option for runtime verification in real-world settings. In this chapter we went through the steps one would take to log event with the necessary data (such as parameters and relevant slices of the system state), and to feed in this runtime data to the verifier. One can find various works exploring its use e.g. [24, 45, 65].

Furthermore, of interest is the compromise of asynchronous monitoring [25] where the monitor is allowed to slack in relation to the system execution but still runs alongside the system, thus possibly still being able to perform reparatory actions when the violation is detected (late).

Chapter 13

A competition aimed at putting different runtime verification tools head-to-head has been run for a number of years [10], with efficiency being one of the main characteristics that a good tool should have. Most papers in the area of runtime verification presenting new tools or variants thereof, report on the efficiency of the tool in question. Amongst the earliest works of runtime verification benchmarking is the work with Tracematches [14]. Since then, several other efforts e.g. [48, 63], have focused on the topic.

There are various distinct directions towards efficient runtime verification apart from simply attempting to synthesise the properties efficiently. A large body of work e.g. [2, 6], focuses on performing static analysis to prune out monitoring parts which can never yield a violation. Another direction is to use statistical sampling to runtime verify only slices of the system behaviour, reaching conclusions with a particular level of confidence [32].

The idea of using runtime verification as testing oracles was implemented in the tool JUnitRV[2] [28]. There are several other interesting ideas for exploiting the close link between runtime verification and testing. One example is the attempt of translating model-based tests into monitors [38], or to have a common specification language from which both tests and monitors can be generated [17]. However, to the best of our knowledge, there are no well established tools which are based on such ideas to date.

The challenges of runtime monitoring distributed system have been studied quite extensively and an overview can be found in [41]. In particular, the architectural design options of orchestration and choreography have been explored in [40] and more recently in [33]. Taking a more general view, a survey of runtime verification challenges can be found in [66].

---

[2] https://www.isp.uni-luebeck.de/junitrv

Chapter 14

A central target of this book is to support software practitioners when introducing runtime verification techniques in industry. A starting point in this regard are [22, 23] analysing experiences of runtime verification in the financial transaction industry. More generally, although sparse, there are other works [7, 67] which look at how runtime monitoring can be incorporated in the software development lifecycle.

References

1. Aceto, L., Achilleos, A., Francalanza, A., Ingólfsdóttir, A., Lehtinen, K.: An operational guide to monitorability with applications to regular properties. Softw. Syst. Model. 20(2), 335–361 (2021), https://doi.org/10.1007/s10270-020-00860-z
2. Ahrendt, W., Chimento, J.M., Pace, G.J., Schneider, G.: Verifying data- and control-oriented properties combining static and runtime verification: theory and tools. Formal Methods Syst. Des. 51(1), 200–265 (2017), https://doi.org/10.1007/s10703-017-0274-y
3. Alur, R., Dill, D.L.: A theory of timed automata. Theor. Comput. Sci. 126(2), 183–235 (1994), https://doi.org/10.1016/0304-3975(94)90010-8
4. Asarin, E., Caspi, P., Maler, O.: Timed regular expressions. J. ACM 49(2), 172–206 (2002), https://doi.org/10.1145/506147.506151
5. Azzopardi, S., Colombo, C., Ebejer, J., Mallia, E., Pace, G.J.: Runtime verification using VALOUR. In: Reger, G., Havelund, K. (eds.) RV-CuBES 2017. An International Workshop on Competitions, Usability, Benchmarks, Evaluation, and Standardisation for Runtime Verification Tools, September 15, 2017, Seattle, WA, USA. Kalpa Publications in Computing, vol. 3, pp. 10–18. EasyChair (2017), https://doi.org/10.29007/bwd4
6. Azzopardi, S., Colombo, C., Pace, G.J.: A model-based approach to combining static and dynamic verification techniques. In: Margaria, T., Steffen, B. (eds.) Leveraging Applications of Formal Methods, Verification and Validation: Foundational Techniques - 7th International Symposium, ISoLA 2016, Imperial, Corfu, Greece, October 10-14, 2016, Proceedings, Part I. Lecture Notes in Computer Science, vol. 9952, pp. 416–430 (2016), https://doi.org/10.1007/978-3-319-47166-2_29
7. Baresi, L., Ghezzi, C.: The disappearing boundary between development-time and run-time. In: Roman, G., Sullivan, K.J. (eds.) Proceedings of the Workshop on Future of Software Engineering Research, FoSER 2010, at the 18th ACM SIGSOFT International Symposium on Foundations of Software Engineering, 2010, Santa Fe, NM, USA, November 7-11, 2010. pp. 17–22. ACM (2010), https://doi.org/10.1145/1882362.1882367
8. Barringer, H., Goldberg, A., Havelund, K., Sen, K.: Rule-based runtime verification. In: Steffen, B., Levi, G. (eds.) Verification, Model Checking, and Abstract Interpretation, 5th International Conference, VMCAI 2004, Venice, Italy, January 11-13, 2004, Proceedings. Lecture Notes in Computer Science, vol. 2937, pp. 44–57. Springer (2004), https://doi.org/10.1007/978-3-540-24622-0_5
9. Barringer, H., Rydeheard, D.E., Havelund, K.: Rule systems for run-time monitoring: from eagle to ruler. J. Log. Comput. 20(3), 675–706 (2010), https://doi.org/10.1093/logcom/exn076

C. Colombo, G. J. Pace, *Runtime Verification*,
https://doi.org/10.1007/978-3-031-09268-8

10. Bartocci, E., Falcone, Y., Bonakdarpour, B., Colombo, C., Decker, N., Havelund, K., Joshi, Y., Klaedtke, F., Milewicz, R., Reger, G., Rosu, G., Signoles, J., Thoma, D., Zalinescu, E., Zhang, Y.: First international competition on runtime verification: rules, benchmarks, tools, and final results of CRV 2014. Int. J. Softw. Tools Technol. Transf. 21(1), 31–70 (2019), https://doi.org/10.1007/s10009-017-0454-5
11. Bartocci, E., Falcone, Y., Francalanza, A., Reger, G.: Introduction to runtime verification. In: Bartocci, E., Falcone, Y. (eds.) Lectures on Runtime Verification - Introductory and Advanced Topics, Lecture Notes in Computer Science, vol. 10457, pp. 1–33. Springer (2018), https://doi.org/10.1007/978-3-319-75632-5_1
12. Bauer, A., Leucker, M., Schallhart, C.: Comparing LTL semantics for runtime verification. J. Log. Comput. 20(3), 651–674 (2010), https://doi.org/10.1093/logcom/exn075
13. Bauer, A., Leucker, M., Schallhart, C.: Runtime verification for LTL and TLTL. ACM Trans. Softw. Eng. Methodol. 20(4), 14:1–14:64 (2011), https://doi.org/10.1145/2000799.2000800
14. Bodden, E., Hendren, L.J., Lam, P., Lhoták, O., Naeem, N.A.: Collaborative runtime verification with tracematches. J. Log. Comput. 20(3), 707–723 (2010), https://doi.org/10.1093/logcom/exn077
15. Brzozowski, J.A.: Derivatives of regular expressions. J. ACM 11(4), 481–494 (1964), https://doi.org/10.1145/321239.321249
16. Bucur, D.: Temporal monitors for tinyos. In: Qadeer, S., Tasiran, S. (eds.) Runtime Verification, Third International Conference, RV 2012, Istanbul, Turkey, September 25-28, 2012, Revised Selected Papers. Lecture Notes in Computer Science, vol. 7687, pp. 96–109. Springer (2012), https://doi.org/10.1007/978-3-642-35632-2_12
17. Cauchi, A., Colombo, C., Francalanza, A., Micallef, M., Pace, G.J.: Using gherkin to extract tests and monitors for safer medical device interaction design. In: Luyten, K., Palanque, P.A. (eds.) Proceedings of the 8th ACM SIGCHI Symposium on Engineering Interactive Computing Systems, EICS 2016, Brussels, Belgium, June 21-24, 2016. pp. 275–280. ACM (2016), https://doi.org/10.1145/2933242.2935868
18. Chaochen, Z., Hoare, C.A.R., Ravn, A.P.: A calculus of durations. Inf. Process. Lett. 40(5), 269–276 (1991), https://doi.org/10.1016/0020-0190(91)90122-X
19. Clarke, L.A., Rosenblum, D.S.: A historical perspective on runtime assertion checking in software development. ACM SIGSOFT Softw. Eng. Notes 31(3), 25–37 (2006), https://doi.org/10.1145/1127878.1127900
20. Colin, S., Mariani, L.: Run-time verification. In: Broy, M., Jonsson, B., Katoen, J., Leucker, M., Pretschner, A. (eds.) Model-Based Testing of Reactive Systems, Advanced Lectures [The volume is the outcome of a research seminar that was held in Schloss Dagstuhl in January 2004]. Lecture Notes in Computer Science, vol. 3472, pp. 525–555. Springer (2004), https://doi.org/10.1007/11498490_24
21. Colombo, C., Francalanza, A., Mizzi, R., Pace, G.J.: polylarva: Runtime verification with configurable resource-aware monitoring boundaries. In: Eleftherakis, G., Hinchey, M., Holcombe, M. (eds.) Software Engineering and Formal Methods - 10th International Conference, SEFM 2012, Thessaloniki, Greece, October 1-5, 2012. Proceedings. Lecture Notes in Computer Science, vol. 7504, pp. 218–232. Springer (2012), https://doi.org/10.1007/978-3-642-33826-7_15
22. Colombo, C., Pace, G.J.: Considering academia-industry projects meta-characteristics in runtime verification design. In: Margaria, T., Steffen, B. (eds.) Leveraging Applications of Formal Methods, Verification and Validation. Industrial Practice - 8th International Symposium, ISoLA 2018, Limassol, Cyprus, November 5-9, 2018, Proceedings, Part IV. Lecture Notes in Computer Science, vol. 11247, pp. 32–41. Springer (2018), https://doi.org/10.1007/978-3-030-03427-6_5
23. Colombo, C., Pace, G.J.: Industrial experiences with runtime verification of financial transaction systems: Lessons learnt and standing challenges. In: Bartocci, E., Falcone, Y. (eds.) Lectures on Runtime Verification - Introductory and Advanced Top-

ics, Lecture Notes in Computer Science, vol. 10457, pp. 211–232. Springer (2018), https://doi.org/10.1007/978-3-319-75632-5_7
24. Colombo, C., Pace, G.J., Abela, P.: Compensation-aware runtime monitoring. In: Barringer, H., Falcone, Y., Finkbeiner, B., Havelund, K., Lee, I., Pace, G.J., Rosu, G., Sokolsky, O., Tillmann, N. (eds.) Runtime Verification - First International Conference, RV 2010, St. Julians, Malta, November 1-4, 2010. Proceedings. Lecture Notes in Computer Science, vol. 6418, pp. 214–228. Springer (2010), https://doi.org/10.1007/978-3-642-16612-9_17
25. Colombo, C., Pace, G.J., Abela, P.: Safer asynchronous runtime monitoring using compensations. Formal Methods Syst. Des. 41(3), 269–294 (2012), https://doi.org/10.1007/s10703-012-0142-8
26. Colombo, C., Pace, G.J., Schneider, G.: LARVA — safer monitoring of real-time java programs (tool paper). In: Hung, D.V., Krishnan, P. (eds.) Seventh IEEE International Conference on Software Engineering and Formal Methods, SEFM 2009, Hanoi, Vietnam, 23-27 November 2009. pp. 33–37. IEEE Computer Society (2009), https://doi.org/10.1109/SEFM.2009.13
27. Colombo, C., Pace, G.J., Schneider, G.: Safe runtime verification of real-time properties. In: Ouaknine, J., Vaandrager, F.W. (eds.) Formal Modeling and Analysis of Timed Systems, 7th International Conference, FORMATS 2009, Budapest, Hungary, September 14-16, 2009. Proceedings. Lecture Notes in Computer Science, vol. 5813, pp. 103–117. Springer (2009), https://doi.org/10.1007/978-3-642-04368-0_10
28. Decker, N., Leucker, M., Thoma, D.: junit^{rv}-adding runtime verification to junit. In: Brat, G., Rungta, N., Venet, A. (eds.) NASA Formal Methods, 5th International Symposium, NFM 2013, Moffett Field, CA, USA, May 14-16, 2013. Proceedings. Lecture Notes in Computer Science, vol. 7871, pp. 459–464. Springer (2013), https://doi.org/10.1007/978-3-642-38088-4_34
29. Delgado, N., Gates, A.Q., Roach, S.: A taxonomy and catalog of runtime software-fault monitoring tools. IEEE Trans. Software Eng. 30(12), 859–872 (2004), https://doi.org/10.1109/TSE.2004.91
30. Dijkstra, E.W.: Letters to the editor: go to statement considered harmful. Commun. ACM 11(3), 147–148 (1968), https://doi.org/10.1145/362929.362947
31. Dijkstra, E.W.: Guarded commands, non-determinancy and a calculus for the derivation of programs. In: Bauer, F.L., Samelson, K. (eds.) Language Hierarchies and Interfaces, International Summer School, Marktoberdorf, Germany, July 23 - August 2, 1975. Lecture Notes in Computer Science, vol. 46, pp. 111–124. Springer (1975), https://doi.org/10.1007/3-540-07994-7_51
32. Dwyer, M.B., Diep, M., Elbaum, S.G.: Reducing the cost of path property monitoring through sampling. In: 23rd IEEE/ACM International Conference on Automated Software Engineering (ASE 2008), 15-19 September 2008, L'Aquila, Italy. pp. 228–237. IEEE Computer Society (2008), https://doi.org/10.1109/ASE.2008.33
33. Falcone, Y.: On decentralized monitoring. In: Nouri, A., Wu, W., Barkaoui, K., Li, Z. (eds.) Verification and Evaluation of Computer and Communication Systems - 15th International Conference, VECoS 2021, Virtual Event, November 22-23, 2021, Revised Selected Papers. Lecture Notes in Computer Science, vol. 13187, pp. 1–16. Springer (2021), https://doi.org/10.1007/978-3-030-98850-0_1
34. Falcone, Y., Fernandez, J., Mounier, L.: What can you verify and enforce at runtime? Int. J. Softw. Tools Technol. Transf. 14(3), 349–382 (2012), https://doi.org/10.1007/s10009-011-0196-8
35. Falcone, Y., Havelund, K., Reger, G.: A tutorial on runtime verification. In: Broy, M., Peled, D.A., Kalus, G. (eds.) Engineering Dependable Software Systems, NATO Science for Peace and Security Series, D: Information and Communication Security, vol. 34, pp. 141–175. IOS Press (2013), https://doi.org/10.3233/978-1-61499-207-3-141

36. Falcone, Y., Krstic, S., Reger, G., Traytel, D.: A taxonomy for classifying runtime verification tools. Int. J. Softw. Tools Technol. Transf. 23(2), 255–284 (2021), https://doi.org/10.1007/s10009-021-00609-z
37. Falcone, Y., Mariani, L., Rollet, A., Saha, S.: Runtime failure prevention and reaction. In: Bartocci, E., Falcone, Y. (eds.) Lectures on Runtime Verification - Introductory and Advanced Topics, Lecture Notes in Computer Science, vol. 10457, pp. 103–134. Springer (2018), https://doi.org/10.1007/978-3-319-75632-5_4
38. Falzon, K., Pace, G.J.: Combining testing and runtime verification techniques. In: Machado, R.J., Maciel, R.S.P., Rubin, J., Botterweck, G. (eds.) Model-Based Methodologies for Pervasive and Embedded Software, 8th International Workshop, MOMPES 2012, Essen, Germany, September 4, 2012. Revised Papers. Lecture Notes in Computer Science, vol. 7706, pp. 38–57. Springer (2012), https://doi.org/10.1007/978-3-642-38209-3_3
39. Floyd, R.W.: Assigning meanings to programs. Proceedings of Symposium on Applied Mathematics 19, 19–32 (1967)
40. Francalanza, A., Gauci, A., Pace, G.J.: Distributed system contract monitoring. J. Log. Algebraic Methods Program. 82(5-7), 186–215 (2013), https://doi.org/10.1016/j.jlap.2013.04.001
41. Francalanza, A., Pérez, J.A., Sánchez, C.: Runtime verification for decentralised and distributed systems. In: Bartocci, E., Falcone, Y. (eds.) Lectures on Runtime Verification - Introductory and Advanced Topics, Lecture Notes in Computer Science, vol. 10457, pp. 176–210. Springer (2018), https://doi.org/10.1007/978-3-319-75632-5_6
42. Giannakopoulou, D., Havelund, K.: Automata-based verification of temporal properties on running programs. In: 16th IEEE International Conference on Automated Software Engineering (ASE 2001), 26-29 November 2001, Coronado Island, San Diego, CA, USA. pp. 412–416. IEEE Computer Society (2001), https://doi.org/10.1109/ASE.2001.989841
43. Goldberg, A., Havelund, K.: Automated runtime verification with eagle. In: Ultes-Nitsche, U., Augusto, J.C., Barjis, J. (eds.) Modelling, Simulation, Verification and Validation of Enterprise Information Systems, Proceedings of the 3rd International Workshop on Modelling, Simulation, Verification and Validation of Enterprise Information Systems, MSVVEIS 2005, In conjunction with ICEIS 2005, Miami, FL, USA, May 2005. INSTICC Press (2005)
44. Gorostiaga, F., Sánchez, C.: Stream runtime verification of real-time event streams with the striver language. Int. J. Softw. Tools Technol. Transf. 23(2), 157–183 (2021), https://doi.org/10.1007/s10009-021-00605-3
45. Havelund, K., Joshi, R.: Experience with rule-based analysis of spacecraft logs. In: Artho, C., Ölveczky, P.C. (eds.) Formal Techniques for Safety-Critical Systems - Third International Workshop, FTSCS 2014, Luxembourg, November 6-7, 2014. Revised Selected Papers. Communications in Computer and Information Science, vol. 476, pp. 1–16. Springer (2014), https://doi.org/10.1007/978-3-319-17581-2_1
46. Hoare, C.A.R.: An axiomatic basis for computer programming. Commun. ACM 12(10), 576–580 (1969), https://doi.org/10.1145/363235.363259
47. Hoare, C.A.R.: Assertions: A personal perspective. IEEE Ann. Hist. Comput. 25(2), 14–25 (2003), https://doi.org/10.1109/MAHC.2003.1203056
48. Javed, O., Binder, W.: Large-scale evaluation of the efficiency of runtime-verification tools in the wild. In: 25th Asia-Pacific Software Engineering Conference, APSEC 2018, Nara, Japan, December 4-7, 2018. pp. 688–692. IEEE (2018), https://doi.org/10.1109/APSEC.2018.00091
49. Jin, D., Meredith, P.O., Lee, C., Rosu, G.: Javamop: Efficient parametric runtime monitoring framework. In: Glinz, M., Murphy, G.C., Pezzè, M. (eds.) 34th International Conference on Software Engineering, ICSE 2012, June 2-9, 2012, Zurich, Switzerland. pp. 1427–1430. IEEE Computer Society (2012), https://doi.org/10.1109/ICSE.2012.6227231

50. Kiczales, G., Hilsdale, E., Hugunin, J., Kersten, M., Palm, J., Griswold, W.G.: An overview of aspectj. In: Knudsen, J.L. (ed.) ECOOP 2001 - Object-Oriented Programming, 15th European Conference, Budapest, Hungary, June 18-22, 2001, Proceedings. Lecture Notes in Computer Science, vol. 2072, pp. 327–353. Springer (2001), https://doi.org/10.1007/3-540-45337-7_18
51. Kiczales, G., Lamping, J., Mendhekar, A., Maeda, C., Lopes, C.V., Loingtier, J., Irwin, J.: Aspect-oriented programming. In: Aksit, M., Matsuoka, S. (eds.) ECOOP'97 - Object-Oriented Programming, 11th European Conference, Jyväskylä, Finland, June 9-13, 1997, Proceedings. Lecture Notes in Computer Science, vol. 1241, pp. 220–242. Springer (1997), https://doi.org/10.1007/BFb0053381
52. Leucker, M., Sánchez, C., Scheffel, T., Schmitz, M., Schramm, A.: Tessla: runtime verification of non-synchronized real-time streams. In: Haddad, H.M., Wainwright, R.L., Chbeir, R. (eds.) Proceedings of the 33rd Annual ACM Symposium on Applied Computing, SAC 2018, Pau, France, April 09-13, 2018. pp. 1925–1933. ACM (2018), https://doi.org/10.1145/3167132.3167338
53. Leucker, M., Schallhart, C.: A brief account of runtime verification. J. Log. Algebr. Program. 78(5), 293–303 (2009), http://dx.doi.org/10.1016/j.jlap.2008.08.004
54. Meredith, P.O., Jin, D., Griffith, D., Chen, F., Rosu, G.: An overview of the MOP runtime verification framework. Int. J. Softw. Tools Technol. Transf. 14(3), 249–289 (2012), https://doi.org/10.1007/s10009-011-0198-6
55. Meyer, B.: Applying "design by contract". Computer 25(10), 40–51 (1992), https://doi.org/10.1109/2.161279
56. Navabpour, S., Bonakdarpour, B., Fischmeister, S.: Path-aware time-triggered runtime verification. In: Qadeer, S., Tasiran, S. (eds.) Runtime Verification, Third International Conference, RV 2012, Istanbul, Turkey, September 25-28, 2012, Revised Selected Papers. Lecture Notes in Computer Science, vol. 7687, pp. 199–213. Springer (2012), https://doi.org/10.1007/978-3-642-35632-2_21
57. Neumann, P.G.: Computer-related risks. Addison-Wesley (1995)
58. Neumann, P.G.: Risks of automation: a cautionary total-system perspective of our cyberfuture. Commun. ACM 59(10), 26–30 (2016), https://doi.org/10.1145/2988445
59. Pinisetty, S., Falcone, Y., Jéron, T., Marchand, H., Rollet, A., Nguena-Timo, O.L.: Runtime enforcement of timed properties. In: Qadeer, S., Tasiran, S. (eds.) Runtime Verification, Third International Conference, RV 2012, Istanbul, Turkey, September 25-28, 2012, Revised Selected Papers. Lecture Notes in Computer Science, vol. 7687, pp. 229–244. Springer (2012), https://doi.org/10.1007/978-3-642-35632-2_23
60. Pnueli, A.: The temporal logic of programs. In: 18th Annual Symposium on Foundations of Computer Science, Providence, Rhode Island, USA, 31 October - 1 November 1977. pp. 46–57. IEEE Computer Society (1977), https://doi.org/10.1109/SFCS.1977.32
61. Reger, G., Cruz, H.C., Rydeheard, D.E.: Marq: Monitoring at runtime with QEA. In: Baier, C., Tinelli, C. (eds.) Tools and Algorithms for the Construction and Analysis of Systems - 21st International Conference, TACAS 2015, Held as Part of the European Joint Conferences on Theory and Practice of Software, ETAPS 2015, London, UK, April 11-18, 2015. Proceedings. Lecture Notes in Computer Science, vol. 9035, pp. 596–610. Springer (2015), https://doi.org/10.1007/978-3-662-46681-0_55
62. Rosu, G., Havelund, K.: Rewriting-based techniques for runtime verification. Autom. Softw. Eng. 12(2), 151–197 (2005), https://doi.org/10.1007/s10515-005-6205-y
63. Rozier, K.Y.: On the evaluation and comparison of runtime verification tools for hardware and cyber-physical systems. In: Reger, G., Havelund, K. (eds.) RV-CuBES 2017. An International Workshop on Competitions, Usability, Benchmarks, Evaluation, and Standardisation for Runtime Verification Tools, September 15, 2017, Seattle, WA, USA. Kalpa Publications in Computing, vol. 3, pp. 123–137. EasyChair (2017), https://doi.org/10.29007/pld3

64. Sammapun, U., Sokolsky, O.: Regular expressions for run-time verification. In: Proceedings of the 1st International Workshop on Automated Technology for Verification and Analysis (ATVA'03) (2003)
65. Sánchez, C.: Online and offline stream runtime verification of synchronous systems. In: Colombo, C., Leucker, M. (eds.) Runtime Verification - 18th International Conference, RV 2018, Limassol, Cyprus, November 10-13, 2018, Proceedings. Lecture Notes in Computer Science, vol. 11237, pp. 138–163. Springer (2018), https://doi.org/10.1007/978-3-030-03769-7_9
66. Sánchez, C., Schneider, G., Ahrendt, W., Bartocci, E., Bianculli, D., Colombo, C., Falcone, Y., Francalanza, A., Krstic, S., Lourenço, J.M., Nickovic, D., Pace, G.J., Rufino, J., Signoles, J., Traytel, D., Weiss, A.: A survey of challenges for runtime verification from advanced application domains (beyond software). CoRR abs/1811.06740 (2018), http://arxiv.org/abs/1811.06740
67. Tamura, G., Villegas, N.M., Müller, H.A., Sousa, J.P., Becker, B., Karsai, G., Mankovski, S., Pezzè, M., Schäfer, W., Tahvildari, L., Wong, K.: Towards practical runtime verification and validation of self-adaptive software systems. In: de Lemos, R., Giese, H., Müller, H.A., Shaw, M. (eds.) Software Engineering for Self-Adaptive Systems II - International Seminar, Dagstuhl Castle, Germany, October 24-29, 2010 Revised Selected and Invited Papers. Lecture Notes in Computer Science, vol. 7475, pp. 108–132. Springer (2010), https://doi.org/10.1007/978-3-642-35813-5_5
68. Ubayashi, N., Tamai, T.: Aspect-oriented programming with model checking. In: Ossher, H., Kiczales, G. (eds.) Proceedings of the 1st International Conference on Aspect-Oriented Software Development, AOSD 2002, University of Twente, Enschede, The Netherlands, April 22-26, 2002. pp. 148–154. ACM (2002), https://doi.org/10.1145/508386.508405
69. Wiener, L.: Digital Woes: Why We Should not Depend on Software. Addison-Wesley (1993)

Index

C. Colombo, G. J. Pace, *Runtime Verification*,
https://doi.org/10.1007/978-3-031-09268-8

Printed in the United States
by Baker & Taylor Publisher Services